Springer Theses

Recognizing Outstanding Ph.D. Research

Aims and Scope

The series "Springer Theses" brings together a selection of the very best Ph.D. theses from around the world and across the physical sciences. Nominated and endorsed by two recognized specialists, each published volume has been selected for its scientific excellence and the high impact of its contents for the pertinent field of research. For greater accessibility to non-specialists, the published versions include an extended introduction, as well as a foreword by the student's supervisor explaining the special relevance of the work for the field. As a whole, the series will provide a valuable resource both for newcomers to the research fields described, and for other scientists seeking detailed background information on special questions. Finally, it provides an accredited documentation of the valuable contributions made by today's younger generation of scientists.

Theses are accepted into the series by invited nomination only and must fulfill all of the following criteria

- They must be written in good English.
- The topic should fall within the confines of Chemistry, Physics, Earth Sciences, Engineering and related interdisciplinary fields such as Materials, Nanoscience, Chemical Engineering, Complex Systems and Biophysics.
- The work reported in the thesis must represent a significant scientific advance.
- If the thesis includes previously published material, permission to reproduce this must be gained from the respective copyright holder.
- They must have been examined and passed during the 12 months prior to nomination.
- Each thesis should include a foreword by the supervisor outlining the significance of its content.
- The theses should have a clearly defined structure including an introduction accessible to scientists not expert in that particular field.

More information about this series at http://www.springer.com/series/8790

Ling Luo

Temporal Modelling of Customer Behaviour

Doctoral Thesis accepted by
The University of Sydney, Australia

Springer

Author
Dr. Ling Luo
School of Computer Science
The University of Sydney
Sydney, NSW, Australia

Supervisors
Prof. Irena Koprinska
School of Computer Science
The University of Sydney
Sydney, NSW, Australia

Dr. Bin Li
Data61, CSIRO
Eveleigh, NSW, Australia

ISSN 2190-5053 ISSN 2190-5061 (electronic)
Springer Theses
ISBN 978-3-030-18288-5 ISBN 978-3-030-18289-2 (eBook)
https://doi.org/10.1007/978-3-030-18289-2

This Springer imprint is published by the registered company Springer Nature Switzerland AG
The registered company address is: Gewerbestrasse 11, 6330 Cham, Switzerland

To my beloved family for their endless support.

Supervisor's Foreword

Understanding user behaviour is important in many practical applications. Machine learning techniques, which can extract useful patterns from vast amounts of collected data, offer promise in providing insights into user behaviour and supporting decision making.

Dr. Ling Luo's Ph.D. thesis is concerned with temporal models of user behaviour. Specifically, Ling has developed machine learning methods based on stochastic processes and temporal collaborative filtering to discover temporal patterns in customer purchase behaviour. One of the main ideas is to use latent variable modelling to segment the customers into groups according to their behaviour and to mine shared purchase patterns, instead of modelling individual behaviours based on sparse and noisy purchase records. This facilitates the tracking of customer behaviour and its evolution over time, including how the groups split and merge, and how individual customers move between these groups. In addition, it helps to improve understanding of customer responses to promotions, health programs and other types of initiatives and interventions.

The proposed methods are evaluated in the context of a health program delivered by an Australian supermarket chain. Participants receive discounts on fruits and vegetables, and have access to a supporting website with interactive and personalised tools, promoting changes in diet and lifestyle. The analysed data includes supermarket purchase transactions, demographic and health information about the participants and activity logs on the program website. The results show the effectiveness of the methods for tracking customer preferences, understanding behaviour changes and identifying the most receptive customers. They can be used to increase the customer engagement and optimise the timing and focus of promotional campaigns and business strategies.

Dr. Luo's methods contribute to the body of knowledge in data mining and behaviour analytics. They are not limited to modelling purchase behaviour and can be applied to analyse other types of user behaviour, e.g. on social networks, educational and recommender systems.

Dr. Ling Luo is an exceptional young scientist. She has already won several awards including the Google Fellowship in Machine Learning, awarded to the "best and brightest" and "most promising young academics in the world". Results based on Ling's Ph.D. work have been published in the proceedings of leading conferences, including the International Joint Conference on Artificial Intelligence (IJCAI 2017), where in addition to being accepted for oral presentation as a full paper, the work was also showcased in the press release opening of the conference as one of two papers "highlighting new and exciting applications of artificial intelligence research".

It has been a great pleasure and privilege working with Ling. I hope you enjoy reading this thesis and find inspiration and value upon its pages.

Sydney, Australia
February 2019

Associate Professor Irena Koprinska

Parts of this thesis have been published in the following articles:

1. Ling Luo, Bin Li, Irena Koprinska, Shlomo Berkovsky, Fang Chen. Who will be affected by supermarket health programs? Tracking customer behaviour changes via preference modelling. In Proceedings of *Pacific Asia Conference on Knowledge Discovery and Data Mining*, pages 527–539. Springer, 2016.

2. Ling Luo, Bin Li, Irena Koprinska, Shlomo Berkovsky, Fang Chen. Discovering temporal purchase patterns with different responses to promotions. In Proceedings of the *25th ACM International Conference on Information and Knowledge Management*, pages 2197–2202. ACM, 2016.

3. Ling Luo, Bin Li, Shlomo Berkovsky, Irena Koprinska, Fang Chen. Online Engagement for a Healthier You: A Case Study of Web-Based Supermarket Health Program. In Companion Proceedings of *ACM World Wide Web Conference*, pages 1053–1061. ACM, 2017.

4. Ling Luo, Bin Li, Irena Koprinska, Shlomo Berkovsky, Fang Chen. Tracking the Evolution of Customer Purchase Behavior Segmentation via a Fragmentation-Coagulation Process. In Proceedings of *International Joint Conference on Artificial Intelligence*, pages 2414–2420. IJCAI, 2017.

Acknowledgements

I would like to thank all the people who have encouraged and supported me to complete my Ph.D. studies and this thesis.

I am sincerely grateful for my primary supervisor Dr. Irena Koprinska, from the School of Computer Science (formerly School of Information Technologies) at the University of Sydney, for her continuous guidance and support throughout these years. When I was an undergraduate student at the University of Sydney, Irena was my lecturer of the Artificial Intelligence course. Her course opened up a fascinating new world for me, and motivated me to pursue a Ph.D. in this direction. During my Ph.D. studies, Irena consistently encouraged and supported me to overcome all kinds of difficulties encountered in my research and my life. She is my role model. I have learned a lot through her commitment to excellence in teaching and doing research.

I would also like to express my most sincere gratitude to my supervisor Dr. Bin Li, who is currently an associate professor at the School of Computer Science, Fudan University. I am very grateful for the huge amount of guidance from Bin at every step along the way, from identifying interesting problems, developing innovative solutions, to writing papers. Bin always provides instant and helpful feedback to my work and inspires me patiently to solve problems. I am extremely fortunate to be Bin's student, and I really enjoy the experience working with him.

I would like to thank Dr. Shlomo Berkovsky, from Data61, for generously sharing datasets and inspiring thoughts with me. I am also thankful to Dr. Fang Chen, from Data61, and Dr. Wei Liu, from University of Technology Sydney, for their great support during my Ph.D. studies.

I would like to give a special thank you to my parents, Yongjun Luo and Yuhuan Ma, for their unreserved love and encouragement all the time. I could feel their strong support no matter where I am. I would also like to thank my husband, Meng Xue, for his love, company, encouragement and understanding at every moment. I genuinely feel blessed to have him by my side.

Contents

1 Introduction 1
- 1.1 Motivation 1
- 1.2 Research Questions 2
- 1.3 Thesis Outline 3
- 1.4 Contributions 5
- References 6

2 Datasets 7
- 2.1 Transaction Records 7
- 2.2 Survey Data 10
- 2.3 Program Website 10

3 Literature Review 15
- 3.1 Temporal Modelling 15
  - 3.1.1 Rule-Based Models 15
  - 3.1.2 Temporal Collaborative Filtering 16
  - 3.1.3 Stochastic Models 18
- 3.2 Customer Segmentation 20
- 3.3 Factors Impacting Customer Behaviour 22
  - 3.3.1 Price and Promotions 22
  - 3.3.2 Other Customers and Advertisements 23
- 3.4 Discussion 24
- References 25

4 Tracking Purchase Behaviour Changes 29
- 4.1 Background 29
- 4.2 Methodology 31
  - 4.2.1 Preliminaries 31
  - 4.2.2 Extract Customer Preferences 33
  - 4.2.3 Construct Temporal Model for Customer Preferences 34

4.2.4 Analyse Customer Preference Changes 38
4.2.5 Evaluate Program Impact on Customer Segments 39
4.3 Case Study 40
4.3.1 Visualisation of Preference Changes 41
4.3.2 Program Impact on Different Types of Customers 43
4.4 Summary 46
References 46

5 Discovering Purchase Behaviour Patterns 49
5.1 Background 49
5.2 Methodology 52
5.2.1 Preliminaries 52
5.2.2 Model Purchase Behaviour via NHPP 55
5.2.3 Segment Customers with Mixture Models 57
5.2.4 Analyse Customer Segments 60
5.3 Case Study 60
5.3.1 Data Preprocessing 61
5.3.2 Fitness of Purchase Events 63
5.3.3 Long-Term and Short-Term Behaviour Patterns 65
5.3.4 Customer Responses to Promotions 70
5.4 Summary 72
References 73

6 Evaluating Impact of the Web-Based Health Program 75
6.1 Background 75
6.2 Methodology 77
6.2.1 Model Individual Purchase Rates 78
6.2.2 Evaluate Program Impact 79
6.3 Case Study 83
6.3.1 Purchase Rate Analysis 83
6.3.2 Distributions of Behaviour Patterns 86
6.3.3 Program Impact Scores 89
6.4 Summary 92
References 93

7 Tracking the Evolution of Customer Segmentations 95
7.1 Background 95
7.2 Methodology 97
7.2.1 Preliminaries 98
7.2.2 Construct Sequence of Customer Partitions 100
7.2.3 Estimate Parameters of FC-CSM 102

7.3 Case Study 106
7.3.1 Fitness of Purchase Rates 107
7.3.2 Impact of Promotions 108
7.3.3 Trajectory of Purchase Behaviour Changes 110
7.4 Summary 115
References 116

8 Conclusions and Future Work 119
8.1 Conclusions 119
8.2 Future Work 121

Author Biography 123

Chapter 1
Introduction

1.1 Motivation

Customer behaviour analysis has been recognised as an indispensable component of business intelligence and marketing [1]. Understanding customer behaviour is of great interest to marketing researchers and business analysts, as this information can help them communicate better with their customers and develop appropriate strategies.

In particular, customer behaviour models allow stakeholders to discover behaviour patterns, track behaviour changes, identify key factors for behaviour changes and monitor a dynamic business environment [2, 3]. Therefore, stakeholders can find the right time to promote products and communicate with their customers [4]. In addition, customer behaviour models with customer segmentation can help businesses to identify the target customers who are most likely to buy their products [5].

In terms of developing business strategies, customer behaviour models are essential for evaluating the impact of marketing campaigns such as sales or health programs. There are various sales promotions aiming to attract customers and create brand loyalty. There is also an increasing number of joint programs run by governments, retailers or other organisations, to raise awareness about particular issues and products, e.g. a health program which motivates people to buy healthier foods. The stakeholders are keen to know if their promotions or programs are effective for different customers, so that they can choose better timing and incentives in the future. With a better understanding of their customers, stakeholders can easily manage the marketing resources and develop cost- and time-efficient strategies, including promotions and tailored programs with social value.

Customer behaviour refers to how customers make decisions about selecting categories and brands, purchasing time and quantity, and using the products. Various external and internal factors can impact the customer behaviour. In more detail, there are two main streams of *external* factors: (1) marketing, which includes price, place and promotion; and (2) environmental, which includes economic, technological and cultural. The *internal* factors affecting customer behaviour often derive from the

L. Luo, *Temporal Modelling of Customer Behaviour*,
Springer Theses, https://doi.org/10.1007/978-3-030-18289-2_1

cultural, social, personal and psychological characteristics of the customers. We are especially interested in the temporal aspects (i.e. purchase time) of the customer behaviour and the important factors associated with customer behaviour changes.

1.2 Research Questions

This thesis focuses on adapting and applying the state-of-the-art data mining and machine learning techniques from the area of temporal modelling to real-world transactional datasets. We aim to find answers to the following questions:

Q1: When do customers buy products?
Q2: How do customers change their purchase behaviour over time?
Q3: What are the temporal patterns of customer purchase behaviour?
Q4: How would different types of customers respond to marketing strategies?

Below, we describe the objective, significance and main tasks for each of these questions.

Q1: When do customers buy products? This is the fundamental question of the temporal analysis of customer behaviour. Constructing an accurate model to capture the purchase time can facilitate answering the other three questions. At the same time, the answers to the other three questions, such as finding the temporal patterns of behaviour (Q3), can improve the estimation of purchase time. In this thesis, we use stochastic models to describe the occurrence of purchase events. These models consider the uncertainty in the purchase decisions as a result of different endogenous and exogenous factors [6]. An accurate behaviour model should have a high fitness to the existing records, and can be easily generalised to future data. Therefore, we design models that can prevent overfitting sparse and noisy individual purchase records by including additional information—the purchase records from similar customers—as a regularisation term. Specifically, we use temporal collaborative filtering (as described in Chap. 4) and mixture modelling (as described in Chaps. 5 and 6), which can not only describe the customer purchase behaviour accurately, but also overcome the overfitting problem given sparse transaction records.

Q2: How do customers change their purchase behaviour over time? and **Q3: What are the temporal patterns of customer purchase behaviour?** We explore these two questions together in this thesis. We construct temporal models based on collaborative filtering and stochastic processes to track the purchase behaviour changes due to various factors. We detect the typical long-term and short-term temporal patterns of behaviour changes. Specifically, the long-term pattern can reveal how the customer preferences drift over time and the impact of seasonal effects on the customer preferences; the short-term pattern can capture the periodic purchase behaviour due to customer needs and regular sales promotions. Characterising the dynamics of purchase behaviour can provide significant information to understand customer needs and analyse customer responses to marketing strategies. Identifying the temporal patterns of behaviour changes is also a compulsory component for an accurate prediction of purchase time [7].

Q4: How would different types of customers respond to marketing strategies? We quantify the purchase behaviour changes of individual customers and compare different types of customers according to their responses to promotions and health programs. We identify latent groups of customers based on their behaviour, where each customer has mixed membership of all latent groups. The individual preferences and purchase behaviour patterns can be derived from the group-level models, weighted by their membership of those groups. We also explore using random partition modelling to track how customer groups evolve over time. Understanding the differences among customers can support businesses to evaluate the impact of their strategies and find valuable customers. It is an essential step towards personalised recommendations and effective target marketing.

These four questions are highly interrelated. We find that discovering temporal patterns can improve the prediction of purchase events, and indicate suitable methods to group customers. In addition, analysing the group-level customer behaviour is helpful in identifying typical behaviour patterns. The group-level modelling has another advantage compared to individual-level modelling—it can overcome overfitting sparse and noisy individual purchase records.

1.3 Thesis Outline

The thesis is organised as follows. Chapter 2 provides background information and introduces the supermarket health program datasets used in this thesis. Chapter 3 reviews the related work on temporal modelling of customer behaviour, customer segmentation and impact factors on customer behaviour. Chapter 3 also identifies and discusses the gaps between the existing work and our models. Chapters 4–7 describe our proposed temporal models of customer behaviour, including for tracking behaviour changes, identifying groups of customers with similar behaviour, discovering behaviour patterns, monitoring the evolution of customer groups and evaluating the impact of discounts and a health program on customer behaviour. Chapter 8 concludes the thesis; we summarise our findings, discuss the implications of our work and suggest future research directions.

More details about our models and case studies described in Chaps. 4–7 are given below.

Chapter 4 focuses on exploring the impact of a health program on purchase preferences for different types of customers. We propose a method based on temporal collaborative filtering to track the customer preference changes induced by the program. Then we compare the program impact on different types of customers, which have been segmented according to their demographic and health information like age, gender, living arrangement and Body Mass Indicator (BMI). Our method:

(1) derives customer preferences from the transaction data;
(2) captures customer behaviour changes via a temporal model;
(3) segments customers based on their demographic and health information;
(4) evaluates the program impact of our supermarket health program on different types of customers.

The results of our case study indicate that, while overall the program had positive effects in encouraging customers to buy healthy food, its impact varied for different customer segments. For example, the program had larger impact on younger customers and obese customers. These results can inform the design of personalised health programs that target specific customers in the future and benefit more people. The publication associated with Chap. 4 is [8].

Chapter 5 focuses on discovering both the customers' long-term purchase patterns, which may be caused by preference changes and seasonal effects, and also the customers' short-term purchase patterns, which may be caused by promotions and personal needs. Since the purchase records of an individual customer may be too sparse to model, we propose to describe the individual behaviour by detecting a number of latent purchase patterns shared by all customers. The latent purchase patterns are modelled via a mixture of Non-Homogeneous Poisson Processes (NHPP), where each Poisson intensity function is composed by a polynomial and a periodic component. Our method:

(1) estimates the occurrence of purchase events more accurately than homogeneous Poisson processes;
(2) discovers the easy-to-interpret long-term gradual changes and short-term periodic changes of purchase behaviour for different customer groups;
(3) identifies the customers who are receptive to promotions using the correlation between behaviour patterns and price information.

Through our case study, we find that the distribution of different long-term patterns is closely related to whether the product is sensitive to the seasonal effects. We also identify some product categories such as ice cream and confectionery, where customers are receptive to promotions. These results are particularly important for optimising the promotion strategies of products from different categories and targeting customers with different behaviour patterns. The publication associated with Chap. 5 is [9].

Chapter 6 explores whether the online interactions with the supporting website of a health program can impact the offline purchase behaviour in supermarkets. We extend the mixture of NHPP from Chap. 5 to model individual purchase behaviour, and compare the purchase behaviour changes of different customers. Our method:

(1) selects different criteria to partition customers based on their levels of interactions with the supporting website;
(2) evaluates quantitatively the impact of the health program and its supporting website on different groups of customers, considering purchase behaviour changes.

Our case study shows that, during the program, the purchase rates of some fresh products increased significantly on average. Given the distribution of behaviour patterns and impact scores, we find that the program imposed a stronger impact on the participants who logged activities and tracked weights. Our method can facilitate the enhancement of personalised health programs, especially those aiming to maximise the program impact through web or mobile applications. The publication associated with Chap. 6 is [10].

Chapter 7 focuses on tracking the evolution of customer segmentation and compares the impact of promotions on different types of customers. We propose a Bayesian nonparametric Customer Segmentation Model based on a Fragmentation-Coagulation process, named FC-CSM, to track the splitting and merging of customer groups. The individual purchase behaviour can be described by group-level purchase rates and the sequence of their group membership. Our method:

(1) achieves high fitness of purchase rates, outperforming models using a mixture of Poisson processes (described in Chaps. 5 and 6);
(2) tracks how the customer groups split and merge and how individual customers change purchase behaviour over time;
(3) evaluates the impact of promotions on the purchase behaviour of different products.

In the case study, our visualisations demonstrate that for the products with high impact of promotions, the customer groups tended to split during promotions, due to different customer responses to the price reductions. The trajectories of individual purchase behaviour can track how customers shift across groups. They also illustrate that the customers with similar total number of purchases may have contrasting purchase patterns. Our model offers a sensitive tool for tracking the group-level and individual-level behaviour changes. This facilitates further understanding of how the promotion campaigns and marketing strategies affect different customers. The publication associated with Chap. 7 is [11].

1.4 Contributions

In summary, the key contributions of this thesis are:

(1) We propose innovative temporal modelling techniques based on collaborative filtering model RMGM-OT and Poisson processes to track customer behaviour changes.
(2) We discover long-term and short-term purchase behaviour patterns due to various factors, such as seasonal effects and drifting customer preferences over time using a mixture of Non-Homogeneous Poisson Processes (NHPP).
(3) We segment customers into groups based on demographic and behavioural attributes, which can help people to identify and compare different types of customers.

(4) We propose a dynamic customer segmentation model FC-CSM, based on Bayesian nonparametric fragmentation-coagulation process to track the evolution of customer groups, including the splitting and merging of groups, and monitor the dynamic customer behaviour at group and individual levels.
(5) We evaluate the impact of: (i) various promotions, (ii) a health program, and (iii) the supporting website of the health program, on different types of customers.

More importantly, our proposed methods contribute to the body of knowledge in data mining and user behaviour analytics. The methods can be easily extended and adapted for modelling other types of behaviour such as online shopping [4], energy usage [12], activities on social networks [13] and educational systems [14].

References

1. Sheth JN, Mittal B, Newman BI (1999) Consumer behavior and beyond. Dryden Press Fort Worth, TX
2. Chen MC, Chiu AL, Chang HH (2005) Mining changes in customer behavior in retail marketing. Expert Syst Appl 28(4):773–781
3. Huang CK, Chang TY, Narayanan BG (2015) Mining the change of customer behavior in dynamic markets. Inf Technol Manage 16(2):117–138
4. Wang J, Zhang Y (2013) Opportunity model for e-commerce recommendation: right product; right time. In: Proceedings of the 36th ACM conference on research and development in information retrieval. ACM, pp 303–312
5. Rossi PE, McCulloch RE, Allenby GM (1996) The value of purchase history data in target marketing. Mark Sci 15(4):321–340
6. Wagner U, Taudes A (1987) Stochastic models of consumer behaviour. Eur J Oper Res 29(1):1–23
7. Kopperschmidt K, Stute W (2009) Purchase timing models in marketing: a review. AStA Adv Stat Anal 93(2):123–149
8. Luo L, Li B, Koprinska I, Berkovsky S, Chen F (2016) Who will be affected by supermarket health programs? Tracking customer behavior changes via preference modeling. In: Pacific Asia conference on knowledge discovery and data mining. Springer, pp 527–539
9. Luo L, Li B, Koprinska I, Berkovsky S, Chen F (2016) Discovering temporal purchase patterns with different responses to promotions. In: Proceedings of the 25th ACM international conference on information and knowledge management. ACM, pp 2197–2202
10. Luo L, Li B, Berkovsky S, Koprinska I, Chen F (2017) Online engagement for a healthier you: a case study of web-based supermarket health program. In: Proceedings of the 26th international conference on world wide web companion. International world wide web conferences steering committee, pp 1053–1061
11. Luo L, Li B, Koprinska I, Berkovsky S, Chen F (2017) Tracking the evolution of customer purchase behavior segmentation via a fragmentation-coagulation process. In: Proceedings of the twenty-sixth international joint conference on artificial intelligence, pp 2414–2420
12. Li L, Zha H (2015) Energy usage behavior modeling in energy disaggregation via marked Hawkes process. In: Proceedings of the 29th AAAI conference on artificial intelligence, pp 672–678
13. Ferraz Costa A, Yamaguchi Y, Juci Machado Traina A, Traina Jr C, Faloutsos C (2015) RSC: Mining and modeling temporal activity in social media. In: Proceedings of the 21th ACM SIGKDD international conference on knowledge discovery and data mining. ACM, pp 269–278
14. Baker RS, Yacef K (2009) The state of educational data mining in 2009: a review and future visions. J Educ Data Min 1(1):3–17

Chapter 2
Datasets

The datasets used in this thesis are from a supermarket health program, named HealthierU, conducted by an Australian national supermarket chain. The program aims to encourage the participants to improve their lifestyle and maintain healthier dietary habits. The duration of the program was 24 weeks, from May to November in 2014, with two cycles of a 12-week diet program. During the program, the participants had access to a supporting website to manage their diet and plan exercises. They also received 10% discount when purchasing fresh fruits and vegetables in the supermarket.

The datasets consist of three components:

- supermarket transaction records;
- surveys of the participants' demographic information and health conditions;
- activity logs of the program's supporting website.

The timelines of the health program and each dataset are illustrated in Fig. 2.1.

2.1 Transaction Records

The transaction records include the purchase data of 931 customers between January 1 and December 31, 2014. Although the health program started in May, the transaction records were collected via supermarket loyalty cards through the whole year, which allows us to compare purchase behaviour before, during and after the program. Each transaction record has a customer ID, product ID, timestamp, product metadata (category, brand, name and bar code), purchased quantity and cost.

The Entity Relationship (ER) diagram for the transaction records is illustrated in Fig. 2.2. In the diagram, the rectangles represent the entities of the dataset; the diamond represents the action; the ovals are the attributes of an entity or action. The attributes that can be used as the identifiers of entities are underlined.

L. Luo, *Temporal Modelling of Customer Behaviour*,
Springer Theses, https://doi.org/10.1007/978-3-030-18289-2_2

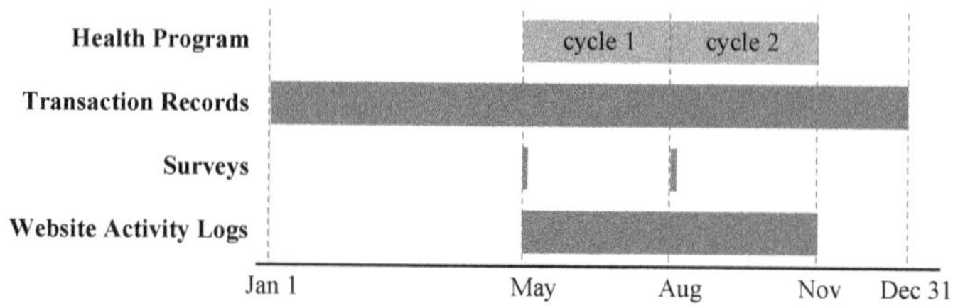

Fig. 2.1 Timelines of the health program and datasets

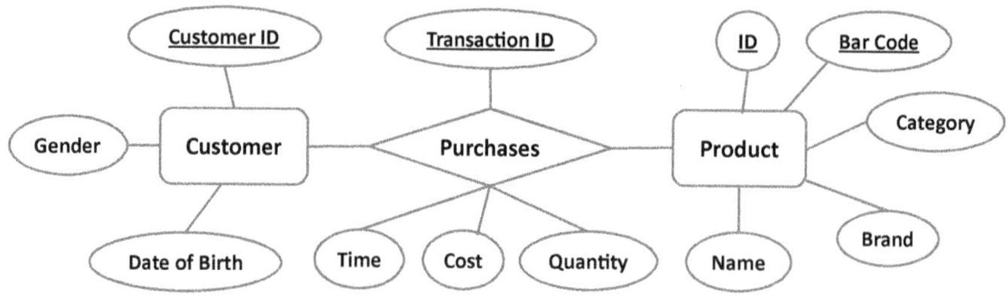

Fig. 2.2 ER diagram for the transaction records

The whole transaction dataset covers over 35,600 products from 202 categories. Given such detailed transaction records, we can analyse the purchase behaviour of these customers from different granularity levels, such as product, brand or category level. There are 110 categories of food and drinks, which are relevant to the health program.

We first explore the number of transactions associated with each category. As shown in Fig. 2.3, we plot the number of transactions for 110 categories of food and drinks, where the categories have been sorted based on the number of transactions in descending order. We notice that about 80% of the transactions are associated with 40 categories. Thus, this plot can provide an overview of the popularity of different categories, and can support us to select popular and representative categories for specific research problems.

To understand the distribution of the transactions and evaluate the activity levels of customers, we count the number of shopping days for each customer. The histogram of the number of shopping days across all customers is shown in Fig. 2.4. About 60% of the customers (556 out of 931) visited the supermarket from 51 to 150 times. The average number of shopping days for all customers is 121.5 days, which means that the customers purchased in the supermarket every 3 days on average.

In order to further examine the overall purchase behaviour, we check the price and daily sales quantity of popular products. For example, Fig. 2.5 shows the price of strawberries and the number of boxes sold on each day. The plot reveals the strong negative correlation between the price and sales quantity of all customers, with a Pearson's correlation coefficient of −0.68. This observation also motivates us

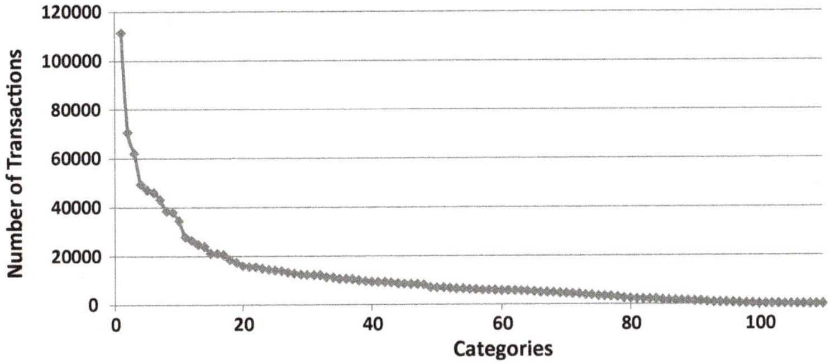

Fig. 2.3 Number of transactions for 110 categories of food and drinks

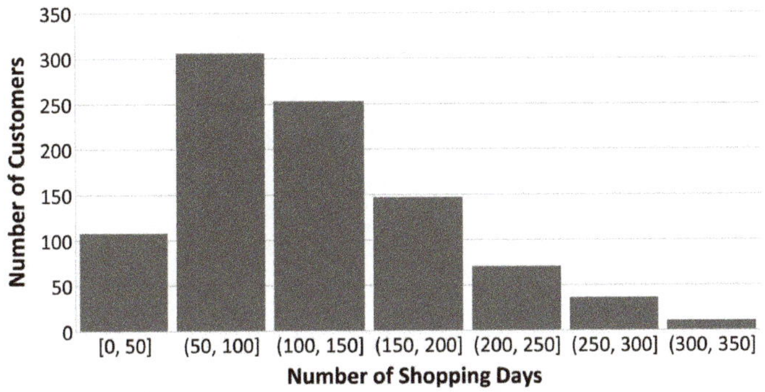

Fig. 2.4 Histogram of the number of shopping days across all customers

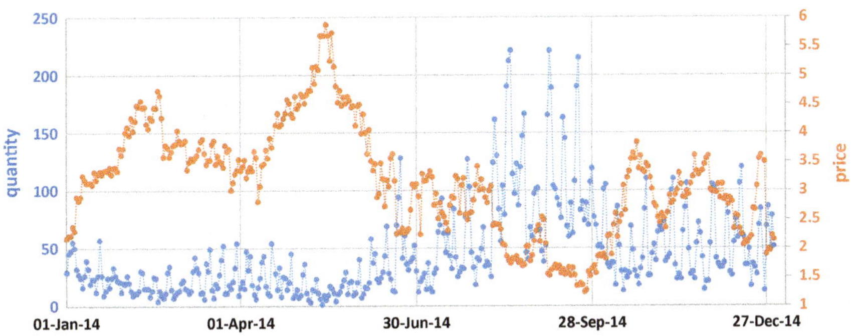

Fig. 2.5 Daily price (orange curve) and number of boxes sold (blue curve) for strawberries

to investigate how different customers would react to price reduction; more details about this are given in Chap. 5.

All of our case studies in Chaps. 4, 5, 6 and 7 involve the transaction records. According to the specific research problems of each case study, we select a subset of the whole transaction records for the study. The details of the data preprocessing for each case study are introduced in their relevant chapter.

2.2 Survey Data

The health program uses surveys to collect the information about the participants, in particular about their demographics, and physical and mental health conditions. The participants were required to complete one survey before each cycle of the 12-week diet program. The first survey was in May and the second was in August.

More precisely, the survey asks questions about the participant's personal details, body measurements, living arrangements, and self-assessment of physical and mental health. It also collects information such as the nutrition and exercise patterns, tobacco and alcohol consumption, personal characteristics, and stress-related life events.

The survey data is mainly involved in our case studies in Chaps. 4 and 6, where we use the survey data to identify different types of participants and examine their engagement with the health program.

2.3 Program Website

The supporting website, HealthierU, provides a comprehensive set of interactive and personalised tools for the diet program, based on the BMI and weight loss goals of customers. The set of tools provided by the website are as follows:

(1) Static program content: overview of the diet program, nutrition tips, meal recipes, instructions for exercises, tutorials on the food groups and guides of the website.
(2) Personalised diet plan: compliant diet plan with recipes tailored to the BMI, food intake, physical activity, weight loss goals and dietary requirements of the user.
(3) Personal diary: the main self-monitoring tool to plan and track daily food intake and exercises (Fig. 2.6).
(4) Weight tracker: the weight recording and progress visualisation tool to track weight changes.
(5) Personal records and results: a set of interactive tools to view the records, measurements and progress towards the weight goals (Figs. 2.7 and 2.8).
(6) Weekly messages: the weekly emails to motivate interactions with the website and a summary of the program.

Overall, the website logged 49,489 actions from all participants over the course of the health program. Each entry contains the user ID, timestamp and the type of the action e.g. visit personal page, view a menu or read a recipe. In Fig. 2.9, we show the weekly total number of actions logged by the program website. We notice that most of the actions took place in the first few weeks of the program. The weekly number

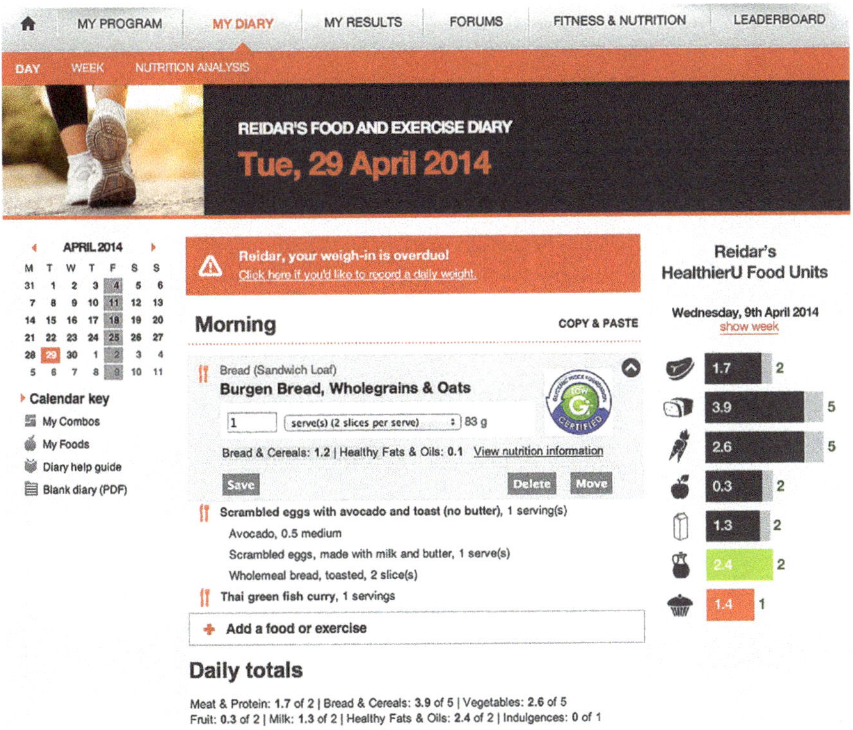

Fig. 2.6 Personal diary page of HealthierU

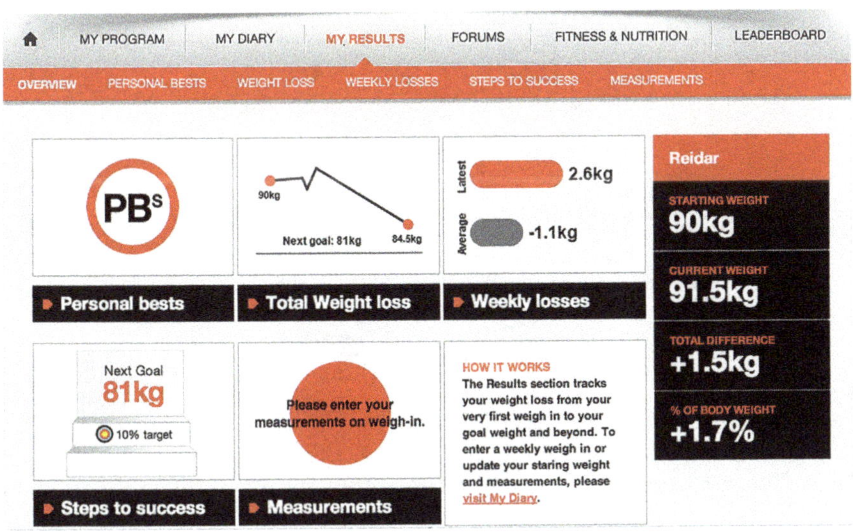

Fig. 2.7 Personal records page of HealthierU

Fig. 2.8 Personal health assessment report of HealthierU

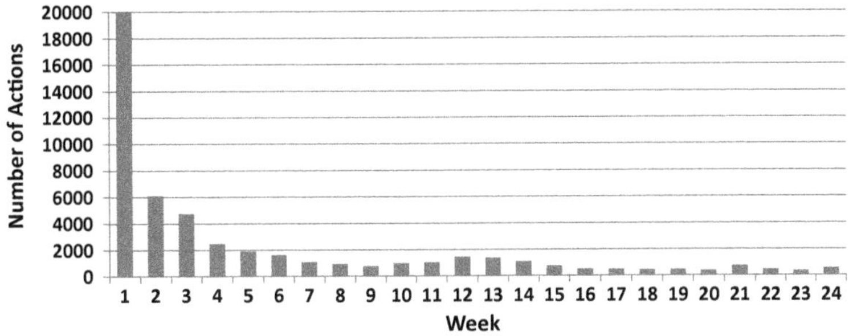

Fig. 2.9 Weekly total number of actions logged by the supporting website during the program

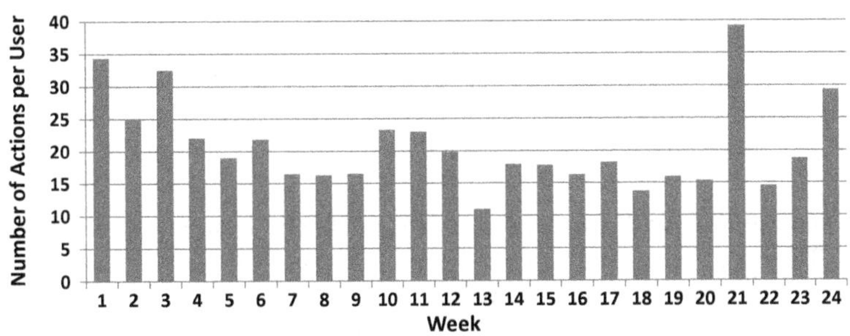

Fig. 2.10 Weekly average number of actions by each web user during the program

of actions remained lower than 2,000 from Week 5, but there was a slight increase in Week 12 and 13 when the second cycle of the program started.

Additionally, we calculate the weekly average number of actions by each website user as shown in Fig. 2.10. Comparing Fig. 2.10 with Fig. 2.9, we find that the average number of actions per website user did not have significantly fluctuation over time, which is different from the case of the weekly total number of actions. Specifically, the total number of actions in Week 1 was about ten times more than the total number of actions in Week 5; while the average number of actions per website user in Week 1 was only two times more than the average number of actions per website user in Week 5. This means that for the users who consistently interacted with the website, their weekly number of actions remained at about 15. Therefore, the main cause of the low level of the website activity is the shrinking number of website users.

Among 32,615 personal diary entries, about 92.5% of the entries were on food and 7.5% were on physical activities. Similar to the website activity, most of the diary entries were written in the first few weeks. We calculate the weekly average number of diaries written by each diary user, as shown in Fig. 2.11. Most of the weekly average numbers of diaries per user ranged from 30 to 70. In both cycles, from Week 1 to Week 12 and from Week 13 to Week 24, the average numbers of diaries per user had increasing trends. This shows that although the number of users who kept

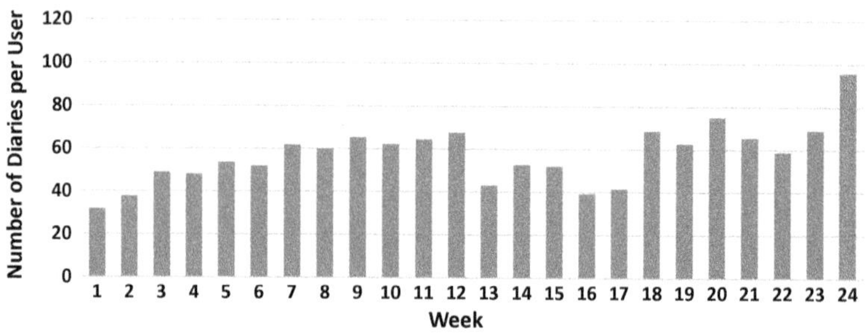

Fig. 2.11 Weekly average number of diaries written by each diary user during the program

diaries decreased throughout each cycle, there was a group of users who increased their diary entries, which indicates that they found the health program useful.

The website usage data is used in Chap. 6, which explores the correlation between participants' engagement with the supporting website and their purchase behaviour changes.

In summary, we use the datasets from a supermarket health program in our case studies. The datasets have three components—transaction records, surveys of participants' basic information, and activity logs of the supporting website. Our case studies in Chaps. 4–7 used one or multiple components of the datasets to model customer behaviour, track their behaviour changes and understand the impact of the health program on different customers.

Chapter 3
Literature Review

As introduced in Chap. 1, our main research tasks are to understand when customers purchase products, detect customer behaviour patterns, identify different types of customers and compare their responses to factors such as promotions and other marketing strategies. Therefore, we review related work from the following areas:

(1) temporal modelling of customer behaviour;
(2) customer segmentation;
(3) factors impacting customer behaviour.

3.1 Temporal Modelling

As customer behaviour often changes over time due to various external and internal factors, constructing accurate temporal models for customer behaviour becomes increasingly important to understand and track customer behaviour changes. The temporal modelling techniques of customer behaviour can be classified into three categories: rule-based models, temporal collaborative filtering and stochastic models.

3.1.1 Rule-Based Models

Rule-based models like association rules [1, 2] and decision trees [3] have been frequently used to model purchase behaviour and capture behaviour changes. There are direct approaches, which compare data from distinct time periods, and incremental approaches, which continuously adjust the model with new transactions.

As for *direct approaches*, Song et al. [2] generated sets of association rules between customer profiles (i.e. demographic information) and products they purchased based on the transaction records at different time snapshots. They measured the similarity between the two sets of rules to detect emerging, added, perished and

L. Luo, *Temporal Modelling of Customer Behaviour*,
Springer Theses, https://doi.org/10.1007/978-3-030-18289-2_3

unexpected change patterns. Chen et al. [1] extended the model in [2] by integrating customer behavioural variables in customer profiles, which used a Recency, Frequency and Monetary (RFM) scoring approach to evaluate the customer value. They also improved the similarity measure to handle multiple attributes in the right-hand-side of the association rules. Huang et al. [4] used fuzzy quantitative sequential patterns to understand the sequence and quantity of products purchased in consecutive shopping trips and identify customer behaviour changes. They developed a sequence similarity computation index to calculate the similarity between patterns for different scenarios. Wang et al. [5] explored the association patterns between pairs of products within or across transactions. Each association pattern had a weight, which represented the strength of the correlation between two products. Their generative processes can help to identify the low-dimensional purchase interests from a large number of association patterns. Liu et al. [3] proposed to build decision trees and identify changes between trees built for different time periods. They compared multiple trees to examine: (1) the partition changes, such as merge or split of nodes, boundary shifts and changes of the leaf node's label, (2) the error rate changes, and (3) the coverage changes.

In terms of *incremental approaches*, Cheung et al. [6] designed a framework called fast update and pruning, which iteratively filtered out redundant rules, generated new frequent itemsets and selected strong rules. They can handle frequent or occasional updates of the database, including adding, deleting or modifying transactions. Masseglia et al. [7] proposed an incremental sequence extraction to discover sequential patterns efficiently when data is changing. Their method can minimise the computational costs by reusing the support of frequent sequences extracted from the earlier datasets. Hong et al. [8] constructed the fast updated frequent-pattern tree (FUFP-tree), which used the incremental tree maintenance algorithm to avoid processing transactions in a batch way. They showed that the FUFP-tree achieved a good trade-off between computational time and tree complexity.

In summary, the main advantage of rule-based methods is that they demonstrate explicitly the relationship between the customer profile and purchase events. The rules produced by the model can be easily explained and used for decision making. The business managers can adapt their strategies and select customer groups in a dynamic environment based on these rules. However, rule-based methods are sensitive to noise and are not suitable for sparse data. Moreover, the previous work only considered a small number of time periods, which can be too coarse-grained to track the gradual behaviour changes over time.

3.1.2 Temporal Collaborative Filtering

Temporal Collaborative Filtering (CF) techniques, which are derived from the area of recommender systems, are effective tools for analysing dynamic user preferences over time. Basically, CF methods estimate the user preference of a product considering the other similar users and products in the system. Therefore, CF methods can reduce the problems due to noise and sparsity of the data to some extent.

The *memory-based CF* methods compute the similarity between users or items based on rating data. Typical methods of this approach include neighbourhood-based CF and user-based top-N recommendation. Ding and Li [9] integrated a weighting scheme in a neighbourhood-based CF approach. Each rating had a time function, and the weight of a rating decreased while the rating got older. They considered both item and user factors in the time function, so the model can provide a personalised decay factor based on individual behaviour. Lathia et al. [10] proposed a temporal CF model, called adaptive-kNN, to iteratively predict user preferences over a growing dataset by automatically assign and update the number of nearest neighbours included in the model. Their empirical analysis showed that the simple adaptive-kNN outperformed other models with static parameters. Xiang et al. [11] proposed the Session-based Temporal Graph (STG), which captured long-term preferences through user-item connections, and short-term preferences through session-item connections. Within the STG framework, they introduced an injected preference fusion to balance the impact of long-term and short-term preferences, and extended the personalised random walk for temporal recommendation. Cho et al. [12] analysed customer purchase behaviour by adapting the self-organising map technique to cluster transactions, identify sequences of clusters and extract sequential rules for clusters. They conducted personalised recommendations based on the extracted behaviour locus for customers.

Instead of explicitly measuring the similarity between users or items, another type of CF methods is based on the *latent factor model*, which identifies a small number of dimensions to represent the original behavioural data. This type of methods can not only reduce the redundant dimensions of the original data space, but also divide items and users into groups with similar preference profiles. The well-known Netflix award-winning method timeSVD++ [13] introduced a time-dependent variable into each user-feature for modelling the temporal dynamics in customer preferences. Their objective function included regularisation terms to describe the time changing baseline predictors—gradual drifting user-bias and item-bias, and dynamic factors derived from factorising the rating matrix for a time interval. TimeSVD++ can effectively distinguish the long-term enduring signals from the transient temporal effects, which is critical in improving the prediction performance. Xiong et al. [14] proposed a factor-based method, Bayesian probabilistic tensor factorisation, for temporal CF. The users, items and time were represented in a shared low-dimensional space, so that the user's preference rating for an item at a certain time can be computed by the inner product of their latent representations. For the time dimension, each time feature vector depended only on the previous time features, which reflected the smooth evolution over time. Their full Bayesian treatment enabled the automatic control of the model complexity and avoided fine parameter tuning.

Yin et al. [15] proposed a dynamic temporal context-aware mixture model to understand evolving user behaviour. They identified the user-oriented latent topics for the user's intrinsic interest and the time-oriented latent topics for the influence of temporal context. The social and temporal correlation information was included as prior knowledge to overcome the problems brought by sparse data.

Additionally, they adopted an item-weighting mechanism to highlight the salient and bursty items while detecting latent topics. Li et al. [16] proposed a dynamic nonlinear Bayesian matrix factorisation to capture the varying rates of preference changes for different customers. They focused on user feature vectors by marginalising out item feature vectors. They considered the previous user features as the prior knowledge in a Bayesian model to ensure that the user preferences change smoothly. Ahmed et al. [17] designed a Bayesian dynamic topic model for time-varying user preferences. They considered the temporal changes in global interests, individual interests and underlying topics. In addition, the streaming distributed inference algorithm allowed applying their model to large-scale datasets.

Li et al. [18] proposed a Rating-Matrix Generative Model Over Time (RMGM-OT), which utilised the advantages of probabilistic topic models to explicitly model the distribution of user preferences over item groups. RMGM-OT can track and visualise how the user preferences drift over time. Moreover, they used two sets of topics for user preferences and item trends, respectively, which had a better interpretability than using a set of shared topics for both factors as the model in [14]. We adapted RMGM-OT to track customer preference changes in Chap. 4, so more technical details about this method are provided in Sect. 4.2.

The temporal CF methods are often designed to use ratings to capture user preferences explicitly. However, the purchase behaviour data is a type of implicit feedback [19], which can introduce bias when it is used in temporal CF. For example, converting purchase records into ratings may distort user preferences, as the purchase quantities are imbalanced, with a lot of ones. In addition, the purchase data cannot reflect negative feedback. Therefore, it requires careful extraction of customer preferences from purchase histories, or adjusting models to handle the purchase quantity or amount of money spent directly.

3.1.3 Stochastic Models

Another line of related work of customer behaviour modelling employs stochastic models to describe the uncertainty of customer behaviour. Ehrenberg [20] proposed to use the Negative Binomial Distribution (NBD) to model the aggregate purchase frequencies of non-durable goods for the population over a period of time. Following Ehrenberg's seminal paper, several extensions [21–24] were proposed to improve the estimation of customers' purchase frequencies.

Chatfield and Goodhardt [21] used the Erlang distribution to model the interval between purchases, rather than the negative exponential distribution in the original NBD model, as they found that the customer purchase behaviour can be more regular than the Poisson process. Morrison et al. [22] presented and compared various NBD mutations within a unifying framework using the conditional expectation of purchase events. They also examined how to generalise NBD to different scenarios, such as allowing for non-buyer and non-stationary behaviour, as the original NBD model did not perform well in these cases.

Wu and Chen [23] proposed to count different types of customers considering critical factors, including the customer's in-store decision, inter-purchase time, repurchasing behaviour and departure. They used a logit model for in-store decisions and the Erlang-c distribution for inter-purchase time. They computed repeat buying probability, which indicated the customer's loyalty. Their integrated model outperformed the other NBD-based models, especially when there were many light buyers. A more recent model in [24] used the log-normal distribution to characterise the heterogeneity of customers, instead of using the Gamma distribution. Trinh et al. [24] showed that the Poisson log-normal model had an intuitive interpretation based on the central limit theorem. The empirical study on multiple categories of a 2-year supermarket dataset showed that their Poisson log-normal model was more accurate than NBD model.

However, these stochastic models focused on estimating the overall number of purchase events of all customers, which is too coarse-grained when it requires analysing behaviour at individual level or group level. Moreover, these models have not considered the temporal dynamics in customer behaviour. The stochastic processes can be used to improve this weakness. Kim et al. [25] proposed a hierarchical time-rescaling model to identify different temporal purchase patterns, such as periodic, bursty, self-exciting and sale-effect patterns, via point processes. They considered individual self-exciting behaviour, marketing stimulus and preceding purchase dependence in a continuous time setting.

The stochastic processes have also been widely applied to different types of user behaviour, such as capturing the temporal patterns of activities on social media. These methods can be easily adapted for customer purchase behaviour. For example, Costa et al. [26] proposed the Rest-Sleep-and-Comment (RSC) model describing the patterns of communication activities on social media platforms. They found that the inter-arrival time of postings had some characteristics, such as positive correlation, periodic spikes, bimodal and heavy-tailed distribution, which violated the assumptions of Poisson processes. Therefore, they used a self-correlated process in RSC to match the patterns observed in real data, and successfully identified bots from real users with a high accuracy. Pan et al. [27] proposed to study trajectory data with marked Poisson processes, where the locations of data points were taken as the marks of the process. The intensity and mark-distribution of the Poisson process were modulated by latent Markov jump processes, which transitioned between multiple states in a continuous time setting. Rao and Teh [28] proposed a generalised Poisson process with a nonparametric prior distribution—Gaussian process—for intensities, which allowed more flexibility when characterising the distribution of the intervals between events. Adams et al. [29] also explored the Poisson process modulated by Gaussian processes. They proposed a tractable inference method for their sigmoidal Gaussian Cox process, which can simulate a Markov chain on the posterior distribution of infinite-dimensional intensity functions, without relying on approximations or finite-dimensional proxy distributions.

Most of the stochastic models above used the Gamma distribution to capture the heterogeneity among customers. This can simplify the inference process, as the Gamma distribution is the conjugate prior for Poisson processes. However, the

Gamma distribution may not be appropriate to capture different behaviour patterns in real cases [24], and it could not explicitly identify different types of customers. For methods working at individual level, such as [25, 26], their performance can be affected by sparse and noisy records of individual customers. Therefore, we also review the related work on segmenting customers, which can be used to enhance the purchase behaviour modelling.

3.2 Customer Segmentation

Various criteria can be used to group customers, including geographic, demographic, psychographic, cultural and behavioural characteristics. For example, given the demographic information of customers, such as gender, age and occupation, it is straightforward for analysts to divide the customers based on these factors, in order to examine the differences among groups, like the number of purchases and response to price reduction.

Dong and Kaiser [30] studied household purchasing and non-purchasing behaviour with the double-hurdle model and marked purchase renewal models. Multiple demographic variables of the households, including income, size, race, age of the head of household and education level, were involved in the models. Their models revealed the impact of different variables on the probability, frequency and amount of their purchase. Taylor et al. [31] developed a scoring system named Healthy Trolley Index to examine dietary quality. They compared the proportions of food expenditure on different product categories with the benchmark provided in the official guide to healthy eating. They grouped different customers by their gender, age and living arrangements. Böttcher et al. [32] proposed to segment customers via the frequent itemset discovery based on the demographic and other attributes of customers. They tracked the growing and shrinking of these segments in dynamic markets and assessed the scores of the detected trends and stabilities.

However, the previous analysis found that it may not be helpful to segment customers based on demographic and psychographic variables for frequently purchased products such as food and drinks [33], whereas using behavioural variables is a sensible approach to segmenting customers [34]. The behavioural segmentation divides customers according to their usage rates, loyalty status, and response to a product. Chen et al. [1] considered behavioural variables when mining association rules from purchase behaviour, in addition to demographic variables and transaction records. They segmented customers based on their values, assessed by the growth matrix of Boston Consulting Group. Considering expenditure and purchase frequency, the customers were classified into four groups: best customers, frequent customers, spenders, and uncertain customers.

Bucklin and Gupta [33] proposed a mixture of logit models to segment customers based on their brand choice and purchase events. The membership of each household was determined by their posterior probabilities in each segment. The segmentation results showed the customer's sensitivities to marketing variables like price and promotions. After segmentation, they also examined the distribution of demographic characteristics within each segment and provided an interpretation for each segment. Bucklin et al. [35] extended the model in [33] by integrating the quantity bought in a purchase event together with the brand choice and purchase event. With similar logit models for the brand choice and purchase event as what had been used in [33], the quantity was modelled by the Poisson distribution. Another main improvement of this work was that they used joint segmentations instead of taking these factors as independent. The joint segmentation can decrease the number of segments significantly without undermining the prediction accuracy.

The latent variable modelling has also been utilised to detect underlying groups for customers and products. Allenby et al. [36] proposed a generalised Gamma component mixture model for the inter-purchase time of individual customers, so that the specific component for an observation was determined by the past behaviour of customers. They introduced random effects through a hierarchical Bayesian structure that yielded the individual-level estimates of key model parameters. They aimed to obtain an early indication of when customers are likely to become inactive. Netzer et al. [37] constructed a non-homogeneous hidden Markov model to capture the dynamics of customer segmentation. They modelled the transitions among latent customer relationship states, such as dormant, transitory and active states, which allowed business analysts to infer the underlying structure of customer relationships. Their model also facilitated business analysts to assess the effects of time-varying covariates on the transition between different states and subsequent purchase behaviour.

Among the latent variable modelling techniques, the topic modelling from text mining, which was originally designed to discover the latent topics in documents, has been adapted to find the latent groups for customer behaviour analysis. Blei and Lafferty [38] described a Dynamic Topic Model (DTM) to track the topics in a sequence of documents, assuming that the topics of these documents will evolve with time. This work extended the previous influential Latent Dirichlet Allocation (LDA) model proposed by Blei et al. [39]. DTM has enabled new ways of understanding large and unstructured document collections. Du et al. in [40] proposed another dynamic model called Sequential LDA. Rather than tracking how topics of a collection of documents change, the Sequential LDA can trace evolving topics within a document such as chapters and paragraphs by hierarchical Poisson-Dirichlet process. Applying the topic modelling idea to customer behaviour analysis, Iwata et al. [41] proposed a topic tracking model, which improved DTM [38], to discover a set of latent topics of customer preferences and the trends of these topics to model the probability of customer purchase behaviour. In order to consider the long-term dependence of customer preferences at the current time interval, their model included a weighted sum of the customer preferences at multiple previous time intervals in the prior distribution. In addition, the probability of a customer being interested in a certain topic can be interpreted as the customer's membership of the group corresponding to that topic.

Christidis et al. [42] used probabilistic topic models to discover latent baskets from transaction records, and find latent user groups according to the series of transactions performed by users. The topic extractions for baskets and customers were based on

LDA. Chung et al. [43] proposed a variational recurrent neural network for sequential data. The method included high-level latent random variables in the hidden state of the neural network to model the variability in sequential data. They also considered the temporal dependencies between latent random variables across subsequent time points. Lloyd et al. [44] proposed a probabilistic topic model to factorise continuous Poisson processes. Each point of the process can be assigned to one of the latent rate functions, and the set of all rate functions were shared across multiple processes. Their model identified sparse and smooth latent factors, which can be applied to customer segmentation.

3.3 Factors Impacting Customer Behaviour

After constructing customer behaviour models, another critical task is to understand various factors that impact customer behaviour. This can help business analysts to evaluate their policies and marketing strategies. We categorised the related work in this area into two groups based on the impact factors: (1) price and promotions and (2) other customers and advertisements.

3.3.1 Price and Promotions

Ball et al. [45] assessed the effects of a 20% price reduction on the purchase and consumption of fruit, vegetables and beverage. They also designed a tailored skill-based behaviour change intervention, which can be combined with the price reduction strategy. Their empirical results showed that the price reduction increased household purchase by 35% for fruit and 15% for vegetables. Their findings indicated that the sustained healthy food subsidies is an effective approach to promoting fruit and vegetable consumption. Iwata and Sawada [46] introduced a topic model for analysing purchase behaviour considering price information. In their model, each topic had a distribution over relevant items and the topic had its own price distribution for each item. The proportion of topics for each user showed the user's purchase preferences, which were determined by the items they bought and the item price. The proposed model can discover the clusters of concurrently purchased items with their price ranges. Adamopoulos and Todri [47] studied the long-term impact of promotions on social media by a causal estimation framework combining econometrics with predictive modelling. They qualified the importance of various factors of promotions, such as discount rate, flexibility and expected benefit. The model allowed businesses to select the optimal type of promotions that can expand the fan base of their brands on social media platforms.

There are several relevant studies exploring how to customise promotions after understanding the impact of price reduction on purchase behaviour. Wan et al. [48] proposed a nested feature-based matrix factorisation framework to model the customer preferences of large-scale grocery shopping data. Their model can estimate price elasticity directly based on the matrix factorisation results. From empirical analysis, they found that price did not impact the decision of category purchase significantly, while it was an important factor in the product choice stage. This indicated that it would be more effective to personalise promotions by targeting appropriate products after knowing the category of customer preferences. Lin et al. [49] designed a real-time promotion framework, which offered a one-time discount to attract a potential customer to make their purchase decision promptly. They proposed two discount-giving strategies: one was an algorithm based on kernel density estimation, which was an effective strategy for offline optimal revenue, and another one was based on the Thompson sampling, which was suitable for online scenarios.

Zhang and Krishnamurthi [50] proposed a decision-support system for customised promotions in online stores. They first constructed a joint purchase behaviour model for purchase time, brand choice and purchase quantity. The model also captured the time-varying behaviour of a household and the heterogeneities among different households, such as variety-seeking or inertia tendency. Based on the model, they inferred the optimal amount of discount on each shopping trip for each household. Zhang and Wedel [51] examined the effectiveness of various promotion strategies in online and offline stores. They compared the expected profit of different promotion strategies at different levels of granularity, including mass market, segment and individual levels. For example, they found that the customised promotions outperformed undifferentiated promotions in online stores at all levels of granularity, while the low redemption rates decreased the effect of customised promotions in offline stores. Therefore, they suggested to use customised promotions in online stores and optimal undifferentiated promotions in offline stores.

3.3.2 Other Customers and Advertisements

In addition to price changes and promotions, customer behaviour can also be affected by other customers through communications and advertisements.

Iwata et al. [52] proposed a shared cascade Poisson process to discover the relations among users. They assumed that an event can trigger a chain of future events. Given the activity sequences of multiple users on each item, they inferred the relations among users shared by all items, the intrinsic influence of each user and the popularity of each item. Their model can facilitate understanding the structure and strength of impact among users and identifying influential users.

Tanaka et al. [53] extended the model in [52] by considering the effects of advertisements at the same time. The mutual triggering effects of their cascade Poisson processes considered user preferences, the influence from other users and the effects of viewing advertisements. Their model allowed segmenting customers based on their sensitivity of these impact factors, and it can track the impact strengths of these triggering factors on individual customers over time. Naik et al. [54] proposed an aggregate advertising response model based on the Kalman filtering to optimise the strategy of advertising scheduling. They tracked the dynamic impact of advertisements over time, considering different factors that could decrease and increase the awareness of advertisements. They found that the impact of advertisements would decrease due to wear-out factors during continuous advertising, but the impact would restore due to forgetting effects during the periods of no advertising. They empirically compared alternative strategies and showed the advantages of using pulsing advertising strategies.

3.4 Discussion

Considering the strengths and weaknesses of previous work, we explore customer behaviour modelling via temporal CF and stochastic models, as they can handle sparse and noisy transaction records better than rule-based models. We notice that the previous temporal CF models mainly relied on explicit ratings, but our transaction datasets only have the timestamp and quantity of each purchase, without explicit ratings for products. Therefore, in Chap. 4, we extract and discretise the customer preferences for products from their purchase records, which fills the gap in previous work and facilitates the application of temporal CF techniques.

To improve the fitness and flexibility of our temporal models, we investigate the application of stochastic processes to customer behaviour modelling. We find that there are no reported methods that could discover long-term trends and short-term periodic behaviour patterns from a collection of sparse transaction records. Hence, in Chap. 5, we propose a mixture of non-homogeneous Poisson processes, with a polynomial and a periodic component in the intensity function, to solve this problem.

Unlike the work in [25] which modelled individual purchase records directly, our mixture model can conduct customer segmentation and identify the purchase patterns of different customer groups. Based on the customer membership of different segments and all segment-wise models, our method in Chap. 6 allows analysing behaviour patterns at both individual level and group level.

As for the group-level behaviour changes, although previous work has considered time-varying groups [37, 41], their dynamic models cannot capture the evolution of customer groups in terms of their splitting and merging. Our dynamic customer segmentation model based on Fragmentation-Coagulation Process in Chap. 7 can fill this gap and provide easy-to-interpret modelling results on the evolution of customer groups.

In terms of the factors impacting customer behaviour, we focus on promotions and a health program, considering the datasets available to us and the context of supermarket purchase behaviour. We find that it is difficult to use the price elasticity metric [48] to track the individual customer's receptiveness to promotions. The reason is that the price elasticity is computed based on the amount of price reduction and purchase increase, which is not available at each time point for us. Therefore, we propose to track the customer's receptiveness to promotions using the correlation between purchase rates and price over a period of time.

References

1. Chen M-C, Chiu A-L, Chang H-H (2005) Mining changes in customer behavior in retail marketing. Expert Syst Appl 28(4):773–781
2. Song HS, Kyeong Kim J, Kim SH (2001) Mining the change of customer behavior in an internet shopping mall. Expert Syst Appl 21(3):157–168
3. Liu B, Hsu W, Han H-S, Xia Y (2000) Mining changes for real-life applications. In: Proceedings of the 2nd international conference on data warehousing and knowledge discovery, vol 1874. Springer, Heidelberg, pp 337–346
4. Huang C-K, Chang T-Y, Narayanan BG (2015) Mining the change of customer behavior in dynamic markets. Inf Technol Manag 16(2):117–138
5. Wang P, Guo J, Lan Y (2014) Modeling retail transaction data for personalized shopping recommendation. In: Proceedings of the 23rd ACM international conference on information and knowledge management, vol 2662020. ACM, pp 1979–1982
6. Cheung DW, Han J, Ng VT, Wong CY (1996) Maintenance of discovered association rules in large databases: an incremental updating technique. In: Proceedings of the 12th international conference on data engineering. IEEE, pp 106–114
7. Masseglia F, Poncelet P, Teisseire M (2003) Incremental mining of sequential patterns in large databases. Data Knowl Eng 46(1):97–121
8. Hong T-P, Lin C-W, Yu-Lung W (2008) Incrementally fast updated frequent pattern trees. Expert Syst Appl 34(4):2424–2435
9. Ding Y, Li X (2005) Time weight collaborative filtering. In: Proceedings of the 14th ACM international conference on information and knowledge management. ACM, pp 485–492
10. Lathia N, Hailes S, Capra L (2009) Temporal collaborative filtering with adaptive neighbourhoods. In: Proceedings of the 32nd international ACM SIGIR conference on research and development in information retrieval. ACM, pp 796–797
11. Xiang L, Yuan Q, Zhao S, Chen L, Zhang X, Yang Q, Sun J (2010) Temporal recommendation on graphs via long- and short-term preference fusion. In: Proceedings of the 16th ACM SIGKDD international conference on knowledge discovery and data mining, vol 1835896. ACM, pp 723–732
12. Cho YB, Cho YH, Kim SH (2005) Mining changes in customer buying behavior for collaborative recommendations. Expert Syst Appl 28(2):359–369
13. Koren Y (2010) Collaborative filtering with temporal dynamics. Commun ACM 53(4):89–97
14. Xiong L, Chen X, Huang T-K, Schneider JG, Carbonell JG (2010) Temporal collaborative filtering with Bayesian probabilistic tensor factorization. In: Proceedings of the 2010 SIAM international conference on data mining, vol 10. SIAM, pp 211–222
15. Yin H, Cui B, Chen L, Zhiting H, Zhou X (2015) Dynamic user modeling in social media systems. ACM Trans Inf Syst 33(3):10
16. Li R, Li B, Jin C, Xue X, Zhu X (2011) Tracking user-preference varying speed in collaborative filtering. In: Proceedings of the 25th AAAI conference on artificial intelligence. AAAI, pp 133–138
17. Ahmed A, Low Y, Aly M, Josifovski V, Smola AJ (2011) Scalable distributed inference of dynamic user interests for behavioral targeting. In: Proceedings of the 17th ACM SIGKDD international conference on knowledge discovery and data mining. ACM, pp 114–122
18. Li B, Zhu X, Li R, Zhang C, Xue X, Wu X (2011) Cross-domain collaborative filtering over time. In: Proceedings of the 22nd international joint conference on artificial intelligence. AAAI Press, pp 2293–2298

19. Hu Y, Koren Y, Volinsky C (2008) Collaborative filtering for implicit feedback datasets. In: Proceedings of the 8th IEEE international conference on data mining. IEEE, pp 263–272
20. Ehrenberg ASC (1959) The pattern of consumer purchases. Appl Stat 26–41
21. Chatfield C, Goodhardt GJ (1973) A consumer purchasing model with Erlang inter-purchase times. J Am Stat Assoc 68(344):828–835
22. Morrison DG, Schmittlein DC (1988) Generalizing the NBD model for customer purchases: what are the implications and is it worth the effort? J Bus Econ Stat 6(2):145–159
23. Couchen W, Chen H-L (2000) Counting your customers: compounding customer's in-store decisions, interpurchase time and repurchasing behavior. Eur J Oper Res 127(1):109–119
24. Trinh G, Rungie C, Wright M, Driesener C, Dawes J (2014) Predicting future purchases with the Poisson log-normal model. Mark Lett 25(2):219–234
25. Kim H, Takaya N, Sawada H (2014) Tracking temporal dynamics of purchase decisions via hierarchical time-rescaling model. In: Proceedings of the 23rd ACM international conference on information and knowledge management. ACM, pp 1389–1398
26. Ferraz Costa A, Yamaguchi Y, Juci Machado Traina A, Traina Jr C, Faloutsos C (2015) RSC: mining and modeling temporal activity in social media. In: Proceedings of the 21th ACM SIGKDD international conference on knowledge discovery and data mining. ACM, pp 269–278
27. Pan J, Rao V, Agarwal PK, Gelfand A (2016) Markov-modulated marked poisson processes for check-in data. In: Proceedings of the 33rd international conference on machine learning, pp 2244–2253
28. Teh YW, Blundell C, Elliott L (2011) Modelling genetic variations using fragmentation-coagulation processes. Adv Neural Inf Process Syst 819–827
29. Adams RP, Murray I, MacKay DJC (2009) Tractable nonparametric Bayesian inference in poisson processes with Gaussian process intensities. In: Proceedings of the 26th international conference on machine learning. ACM, pp 9–16
30. Dong D, Kaiser HM (2008) Studying household purchasing and nonpurchasing behaviour for a frequently consumed commodity: two models. Appl Econ 40(15):1941–1951
31. Taylor A, Wilson F, Hendrie G, Allman-Farinelli M, Noakes M (2015) Feasibility of a healthy trolley index to assess dietary quality of the household food supply. Br J Nutr 114(12):2129–2137
32. Böttcher M, Spott M, Nauck D, Kruse R (2009) Mining changing customer segments in dynamic markets. Expert Syst Appl 36(1):155–164
33. Bucklin RE, Gupta S (1992) Brand choice, purchase incidence, and segmentation: an integrated modeling approach. J Mark Res
34. Kotler P, Armstrong G (2010) Principles of marketing. Pearson Education, UK
35. Bucklin RE, Gupta S, Siddarth S (1998) Determining segmentation in sales response across consumer purchase behaviors. J Mark Res 189–197
36. Allenby GM, Leone RP, Jen L (1999) A dynamic model of purchase timing with application to direct marketing. J Am Stat Assoc 94(446):365–374
37. Netzer O, Lattin JM, Srinivasan V (2008) A hidden Markov model of customer relationship dynamics. Mark Sci 27(2):185–204
38. Blei DM, Lafferty JD (2006) Dynamic topic models. In: Proceedings of the 23rd international conference on machine learning. ACM, pp 113–120
39. Blei DM, Ng AY, Jordan MI (2003) Latent dirichlet allocation. J Mach Learn Res 3:993–1022
40. Lan D, Buntine W, Jin H, Chen C (2012) Sequential latent dirichlet allocation. Knowl Inf Syst 31(3):475–503
41. Iwata T, Watanabe S, Yamada T, Ueda N (2009) Topic tracking model for analyzing consumer purchase behavior. In: Proceedings of the 22nd international joint conference on artificial intelligence. AAAI Press, pp 1427–1432
42. Christidis K, Apostolou D, Mentzas G (2010) Exploring customer preferences with probabilistic topic models. In: Proceedings of the joint European conference on machine learning and principles and practice of knowledge discovery in databases

43. Chung J, Kastner K, Dinh L, Goel K, Courville AC, Bengio Y (2015) A recurrent latent variable model for sequential data. In: Advances in neural information processing systems, pp 2980–2988
44. Lloyd C, Gunter T, Nickson T, Osborne MA, Roberts SJ (2016) Latent poisson process allocation. In: Proceedings of the 19th international conference on artificial intelligence and statistics, pp 389–397
45. Ball K, McNaughton SA, Le HND, Gold L, Ni Mhurchu C, Abbott G, Pollard C, Crawford D (2015) Influence of price discounts and skill-building strategies on purchase and consumption of healthy food and beverages: outcomes of the supermarket healthy eating for life randomized controlled trial. Am J Clin Nutr 101(5):1055–1064
46. Iwata T, Sawada H (2013) Topic model for analyzing purchase data with price information. Data Min Knowl Discov 26(3):559–573
47. Adamopoulos P, Todri V (2015) The effectiveness of marketing strategies in social media: evidence from promotional events. In: Proceedings of the 21th ACM SIGKDD international conference on knowledge discovery and data mining. ACM, pp 1641–1650
48. Wan M, Wang D, Goldman M, Taddy M, Rao J, Liu J, Lymberopoulos D, McAuley J (2017) Modeling consumer preferences and price sensitivities from large-scale grocery shopping transaction logs. In: Proceedings of the 26th international conference on world wide web. International World Wide Web Conferences Steering Committee, pp 1103–1112
49. Lin Y-C, Huang C-H, Hsieh C-C, Shu Y-C, Chuang K-T (2017) Monetary discount strategies for real-time promotion campaign. In: Proceedings of the 26th international conference on world wide web. International World Wide Web Conferences Steering Committee, pp 1123–1132
50. Zhang J, Krishnamurthi L (2004) Customizing promotions in online stores. Mark Sci 23(4):561–578
51. Zhang J, Wedel M (2009) The effectiveness of customized promotions in online and offline stores. J Mark Res 46(2):190–206
52. Iwata T, Shah A, Ghahramani Z (2013) Discovering latent influence in online social activities via shared cascade poisson processes. In: Proceedings of the 19th ACM SIGKDD international conference on knowledge discovery and data mining. ACM, pp 266–274
53. Tanaka Y, Kurashima T, Fujiwara Y, Iwata T, Sawada H (2016) Inferring latent triggers of purchases with consideration of social effects and media advertisements. In: Proceedings of the 9th ACM international conference on web search and data mining. ACM, pp 543–552
54. Naik PA, Mantrala MK, Sawyer AG (1998) Planning media schedules in the presence of dynamic advertising quality. Mark Sci 17(3):214–235

Chapter 4
Tracking Purchase Behaviour Changes

4.1 Background

The World Health Organisation (WHO) reports that in 2014 more than 1.9 billion adults were overweight, and over 600 million were obese [1]. Being overweight or obese increases the risk of cardiovascular problems, diabetes and musculoskeletal disorders. To address the obesity problem, numerous behaviour change programs have been designed, aiming at encouraging participants to maintain a healthier lifestyle, e.g. change their diet and perform physical activities [2, 3]. The program stakeholders often desire to know to what extent their program influences the participants, and more importantly, how to improve the program to benefit a broader population in the future. Therefore, mining the vast amount of participant behaviour data collected by the program and understanding their behaviour changes have become critical and timely research tasks.

Problem Formulation and Organisation

For our supermarket health program, we would like to explore in depth how the participant behaviour changes over time and investigate how different types of participants are affected by the health program. Based on the purchase behaviour observed before and after joining the health program, we propose a systematic approach for tracking the customer behaviour changes induced by the health program. We evaluate how the customers from different segments, formed by demographic and health information, are influenced by the health program.

Our method comprises four specialised modules: (1) extracting customer preferences from their transaction data, (2) constructing temporal preference models, (3) analysing behaviour changes since joining the health program, (4) most importantly, quantifying the program impact on different types of customers.

This chapter focuses on exploring the impact of a health program on purchase preferences for different types of participants. The chapter is organised as follows. This section introduces the background, main contributions and related work of the chapter. Section 4.2 firstly describes the preliminaries, including collaborative filtering based on latent variable models and Gibbs sampling, which are used in the model construction. Then the section explains the four specialised modules of

L. Luo, *Temporal Modelling of Customer Behaviour*,
Springer Theses, https://doi.org/10.1007/978-3-030-18289-2_4

our method in detail. Section 4.3 presents our case study on the supermarket health program and analyses the case study results. Section 4.4 summarises the findings and contributions of this chapter.

Contributions

The main contributions of this chapter are listed as follows:

- We construct a temporal preference model based on collaborative filtering that tracks and visualises the preference changes of the program participants over time.
- We quantify the customer preference changes as well as the program impact on different types of customers. The results show that female customers, younger customers, customers who live with their family, and obese customers are more likely to be influenced by the health program.
- Our method and results can potentially facilitate effective personalised health programs in the future.

Related Work

A health program is a specific type of campaign, which promotes a healthier lifestyle and behaviour [2]. The impact of a program is usually evaluated by investigating customer behaviour changes, which can be categorised into two groups:

(1) *direct* approaches, that model the behaviour data at different time periods to identify differences [4, 5];
(2) *incremental* approaches, that continuously adjusted the model with new transactions [6]. For example, the rule-based methods such as association rules [4] and decision trees [5] are frequently used due to their easy interpretation.

However, the direct approaches can be too coarse-grained to reflect the temporal dynamics, while the incremental approaches are sensitive to noise.

Temporal Collaborative Filtering (CF) techniques are powerful tools for analysing the patterns of customer preferences over time. Koren [7] proposed timeSVD++, which integrated a time-dependent factor into each user-feature for modelling time-varying customer preferences for items. In [8], preference changes were analysed via Bayesian tensor factorisation, where the users, items and time can be represented in a shared low-dimensional space. Fanaee-T and Gama [9] proposed a histogram-based Multi-Aspect-Streaming Tensor Analysis model (MASTA), which can avoid the scalability issues of traditional tensor decomposition approaches when tracking changes or detecting patterns at each mode of the data. However, MASTA is limited to analysis tasks, without estimating missing values inside the tensor. Another weakness of MASTA is that it could not handle very dense or very sparse data. In contrast, Rating-Matrix Generative Model Over Time (RMGM-OT) [10] took the advantages of probabilistic topic models to explicitly model the distributions of customer preference over item groups; in addition, the method can illustrate how customer preferences drift over time.

Our goal is to design a model that facilitates a direct comparison across multiple time periods, instead of just before and after the program. Therefore, our customer behaviour model adapts the temporal CF technique – RMGM-OT, to track and visualise the changes of customer preference. We also design procedures to infer implicit customer preferences based on the transaction records, which facilitates the application of temporal CF to purchase behaviour analysis.

4.2 Methodology

In this section, we first introduce the CF based on latent variable models and Gibbs sampling, which are preliminaries of our model. Then we describe our proposed method for measuring the impact of the health program.

As shown in the flowchart (Fig. 4.1), our method consists of four modules:

(1) extracting customer preferences from the transaction data;
(2) constructing a temporal model for customer preferences and visualising customer preferences;
(3) analysing preference changes over time;
(4) evaluating program impact on different types of customers and provide program analytics as output.

The detail of these modules will be provided after Sect. 4.2.1.

4.2.1 Preliminaries

Collaborative Filtering Based on Latent Variable Models

CF is a popular and effective technique in recommender systems [11]. CF can predict the behaviour or preference of a user based on the other users in the system. Basically, it assumes that if users have similar preference for some items, they are likely to have similar preference for other items. Therefore, CF can be employed to estimate the user preferences and perform recommendations for users.

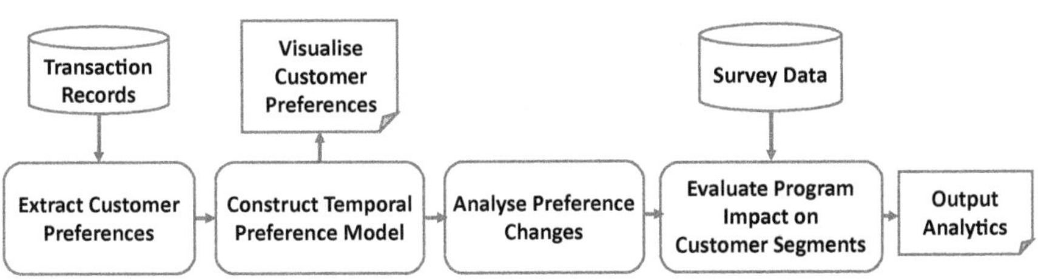

Fig. 4.1 Flowchart of our method

There are mainly three types of CF approaches: (1) the memory-based, (2) the model-based, and (3) the hybrid methods, which integrate both memory-based and model-based CF methods.

For the memory-based methods, they can be further split into two subgroups – the user-based and item-based methods. Assuming we are modelling user u_i's preference for item v_j, a user-based method would first compute the similarity between users based on their preferences (e.g. rating data), and identify users who are similar to u_i with respect to their preferences for other items. Then, the method uses those users' preference for v_j to predict u_i's preference for v_j. The item-based method searches for items that are similar to v_j, and uses u_i's preference for those items to predict u_i's preference for v_j. However, since it is difficult and inaccurate to compute similarity between users (or items) with many missing values, the performance of memory-based methods decreases on sparse dataset.

The model-based methods use machine learning and data mining algorithms to discover the patterns in training data and make predictions. Some model-based methods work on reducing the dimensionality of the user-item preference rating space, and divide users and items into a number of groups with similar preference profile [11]. For example, the Singular Value Decomposition (SVD) can be used to identify k topics based on rating data, so that the user preferences for items can be represented by the user's interest in a topic and the item's relevance to the topic [12]. Another example is the probabilistic Latent Semantic Analysis (pLSA) [13], a technique from information retrieval, which has been successfully applied to CF [14]. Additionally, some methods, such as Bi-Latent Dirichlet Allocation (Bi-LDA) [15], use two sets of latent variables, which are for latent user groups and latent item groups, respectively.

In more detail, Bi-LDA is a latent variable model for matrix tri-factorisation, which consists of two interacting LDA models. The basic idea of applying Bi-LDA to CF is: given a preference matrix $\boldsymbol{X}$ with N users and M items, Bi-LDA infers the latent variables (z^U, z^I) for each user-item pair based on their preference rating. The method factorises $\boldsymbol{X}$ into three matrices $\boldsymbol{P}$, $\boldsymbol{B}$ and $\boldsymbol{Q}$, where $\boldsymbol{B}$ is a low-rank representation of $\boldsymbol{X}$. Matrix $\boldsymbol{B}$ can be interpreted as the preferences of K latent user groups for L latent item groups. Matrices $\boldsymbol{P}$ and $\boldsymbol{Q}$ contain the latent-group membership for individual users and items. Each user u_i has a mixed membership over all user groups, so the element p_{ik} in matrix $\boldsymbol{P}$ denotes u_i's membership in group k, and $\sum_k p_{ik} = 1$. It is similar for item membership in $\boldsymbol{Q}$.

The generative process of Bi-LDA is as follows:

(1) For user-item joint group (k, l), draw $\phi_{kl} \sim Dirichlet(\beta)$;
(2) For user u_i, draw $\boldsymbol{p}_i \sim Dirichlet(\alpha^U)$;
(3) For item v_j, draw $\boldsymbol{q}_j \sim Dirichlet(\alpha^I)$;
(4) For preference rating x_{ij}:
draw a user group $z_{ij}^U \sim Multinomial(\boldsymbol{p}_i)$;
draw an item group $z_{ij}^I \sim Multinomial(\boldsymbol{q}_j)$;
draw a rating $x_{ij} \sim Multinomial(\phi_{z_{ij}^U z_{ij}^I})$.

where α^U, α^I and β are hyperparameters of the Dirichlet priors; ϕ_{kl} is the distribution of preference levels ($\{1, \ldots, R\}$) for a user-item joint group (k, l).

RMGM-OT [10] is a temporal CF technique, which extends Bi-LDA to deal with the temporal information associated with user preferences. RMGM-OT can effectively track and visualise how user preferences drift over time, so it acts as the base pinning for our customer preference model.

Gibbs Sampling

Gibbs sampling [16] is a widely applicable Markov Chain Monte Carlo (MCMC) method [17] to generate a sequence of approximate observations, when it is difficult to sample from that distribution directly. Basically, MCMC method constructs a Markov chain, whose equilibrium distribution is the probability distribution we would like to sample from. After the chain reaches the stable state, each transition in the chain can generate one sample from the desired distribution.

Gibbs sampling is particularly helpful for approximate inference in high-dimensional models. We use Gibbs sampling to infer the parameters of our model. Generally, in order to sample from a distribution with n variables, $p(\boldsymbol{x}) = p(x_1, \ldots, x_n)$, Gibbs sampling draws x_i from the conditional probability $p(x_i|\boldsymbol{x}^{\neg i})$, where $\boldsymbol{x}^{\neg i}$ refers to all variables in $\boldsymbol{x}$ except x_i. In one sampling round, the algorithm visits all variables $\{x_1, \ldots, x_n\}$ randomly or in a particular order and updates them based on the sampling result [18]. The algorithm has a burn-in period before reaching the stable state, which often involves hundreds of sampling rounds. The samples generated during the burn-in period will be discarded. After reaching the stable state, the samples can be used to approximate the desired distribution.

4.2.2 Extract Customer Preferences

The first module extracts customer preferences from the transaction data. Two types of the program data are involved in this study, which are:

(1) The Purchase records of customers captured through loyalty cards between January 1 and December 31;
(2) Self-reported survey data addressing their demographic and health information.

As we are interested in food and drinks that are relevant to the health program, the dataset was reduced to 3,394 products from 24 categories, ranging from fresh products (e.g. vegetables and fruits) to snacks and soft drinks. We selected 884 out of 931 customers, who had at least 50 purchase records of different products from these 24 categories.

Formally, given the transaction dataset comprising N customers and M products, each transaction record can be represented by a tuple (u_i, v_j, q, t), which indicates that customer u_i purchased quantity q of product v_j at time t. Based on all transaction tuples, we construct an $N \times M \times T$ customer preference tensor $\boldsymbol{\mathcal{X}}$, where each element $x_{ij}^{(t)}$ denotes u_i's preference for v_j in R levels ($\{1, \ldots, R\}$) at $t \in \{1, \ldots, T\}$.

The whole time span of the transaction data is evenly partitioned into T periods, and the time index t is determined by the period into which *purchase time* falls. For example, if each time period corresponds to one month, $t = 1$ refers to January. In this way, we slice $\mathcal{X}$ into T preference matrices $\{\boldsymbol{X}^{(1)}, \ldots, \boldsymbol{X}^{(T)}\}$, corresponding to T periods. Each $\boldsymbol{X}^{(t)}$ contains the preference information of N customers for M products in time period t. The series of preference matrices $\{\boldsymbol{X}^{(1)}, \ldots, \boldsymbol{X}^{(T)}\}$ can inform the temporal analysis of customer preference changes.

In our study, the transaction data is partitioned monthly, so the dimension of $\mathcal{X}$ is 844 customers $\times$ 3, 394 products $\times$ 12 months. Each element $x_{ij}^{(t)}$ is a discrete preference value derived from the aggregated amount of product v_j bought by customer u_i in month t. We use discrete preference values rather than actual purchase quantities in $\mathcal{X}$, because the purchase quantities for different products can be incomparable. For example, three apples and three watermelons may not reflect the same level of preference for these two products. Therefore, the preference level is defined according to the rank of a purchase quantity among the purchase quantities of all customers. For a certain product, the customer's preference level is determined by their rank among all customers buying that product.

In more detail, we count the total amount of v_j bought by u_i in period t. Then, we sort the purchase quantities of v_j of all N customers in period t in ascending order and split them into 5 quantiles. The purchase quantities are discretised into 5 levels — $\{1, 2, 3, 4, 5\}$. For example, the values falling in the first 5-quantile (i.e. ranked within the first 20%) become 1 — the lowest preference level, and the values greater than the first 5-quantile but smaller than the second 5-quantile (i.e. ranked within 20% – 40%) becomes 2, and so on. Particularly, if u_i did not purchase v_j in month t, then $x_{ij}^{(t)}$ is a missing value.

4.2.3 Construct Temporal Model for Customer Preferences

The second module is to construct temporal model for customer preferences, which is the key component in order to track customer purchase behaviour changes and investigate how the health program influences their behaviour. However, we notice that if we analyse purchase behaviour based on the product-level preference matrix $\boldsymbol{X}^{(t)}$ directly, the matrix can be very sparse in real-life cases, with a lot of missing values. It is important to note that the missing values do not necessarily mean the lowest preference level, because it is common for customers not to buy certain products at certain periods. Therefore, the temporal CF techniques can be exploited to estimate and smooth customer preferences over time.

In our study, we are particularly interested in the customer preference changes at category level, rather than at individual product level. Considering the above requirements, we adapt the temporal CF method RMGM-OT [10] to our problem setting to capture the temporal dynamics of the customer preference for product categories.

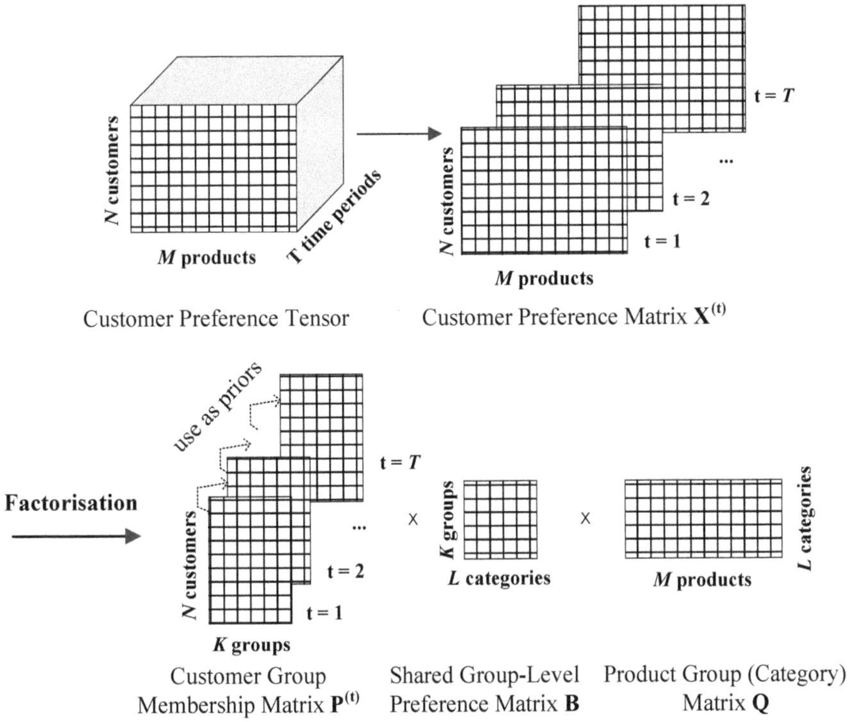

Fig. 4.2 Illustration of factorising customer preference tensor $\mathcal{X}$

Specifically, the preference matrix $\boldsymbol{X}$ can be factorised into three matrices, $\hat{\boldsymbol{X}} = \boldsymbol{P}\boldsymbol{B}\boldsymbol{Q}^\top$, as schematically shown in Fig. 4.2. In the setting of customer preference analysis, the above factorisation can co-cluster customers and products, which results in K latent customer groups and L latent product groups.

The definition of each matrix is as follows:

(1) $\boldsymbol{P}$ is an $N \times K$ customer group membership matrix. Each row $\boldsymbol{p}_i$ can be interpreted as u_i's soft membership distribution over K customer groups, and $\sum_k p_{ik} = 1$ ($p_{ik} \in [0, 1]$).
(2) $\boldsymbol{B}$ is a $K \times L$ group-level preference matrix, which represents the preferences of K customer groups for L product groups.
(3) $\boldsymbol{Q}$ is an $M \times L$ product group membership matrix. It contains the membership information of M products over L product groups. Each column $\boldsymbol{q}_j$ is the category of product v_j, and $q_{jl} \in \{0, 1\}$.

It is worth noting that, in our study, we adopt predefined categories of supermarket products, e.g. vegetables, fruits and soft drinks, to define the product groups. Thus, v_j belongs to only one product category: $q_{jl} = 1$ if l is the predefined category; otherwise, $q_{jl} = 0$.

By taking the temporal domain into consideration, the varying customer preference over time can be modelled via the series of preference matrices $\{\boldsymbol{X}^{(1)}, \ldots, \boldsymbol{X}^{(T)}\}$. We can obtain the customer-group membership matrix $\boldsymbol{P}^{(t)}$ for each time period t. In addition, we assume that the customer-group membership satisfies the Markov property, i.e. the state at t depends on the previous state at $t-1$. Considering the empirical Bayes' approach, we can simply use the preceding customer-group membership $\boldsymbol{P}^{(t-1)}$, weighted by λ as the prior knowledge of the current customer-group membership $\boldsymbol{P}^{(t)}$ in the model (illustrated by the dotted arrows in Fig. 4.2). In terms of the group-level preference matrix $\boldsymbol{B}$, we assume that the group-level preferences are stable across time periods, so $\boldsymbol{B}$ is *shared by all time periods*. The product-group membership matrix $\boldsymbol{Q}$ is predefined based on the product information, so it remains unchanged across time periods.

From the perspective of generative models, the ratings in preference matrix $\boldsymbol{X}^{(t)}$ are generated in the following steps:

(1) For customer-product joint group (k, l), draw $\phi_{kl}^{(t)} \sim Dirichlet(\beta)$;
(2) For customer u_i:
  at $t = 1$, draw $\boldsymbol{p}_i^{(t)} \sim Dirichlet(\alpha^U)$;
  at $t = \{2, \ldots, T\}$, draw $\boldsymbol{p}_i^{(t)} \sim Dirichlet(\lambda \boldsymbol{p}_i^{(t-1)})$;
(3) For product v_j:
  $q_{jl}^{(t)} = 1$ for the predefined category l;
  $q_{jl}^{(t)} = 0$ for other categories;
(4) For preference rating $x_{ij}^{(t)}$:
  draw a customer group $z_{ij}^{U(t)} \sim Multinomial(\boldsymbol{p}_i^{(t)})$;
  draw a rating $x_{ij}^{(t)} \sim Multinomial(\phi_{z_{ij}^U l}^{(t)})$.

where α^U and β are hyperparameters; λ is the weight of the prior knowledge, which controls the dependency between $\boldsymbol{P}^{(t-1)}$ and $\boldsymbol{P}^{(t)}$. The prior distribution of $\boldsymbol{p}_i^{(t)}$ is $Dirichlet(\alpha^U)$ at $t = 1$ and $Dirichlet(\lambda \boldsymbol{p}_i^{(t-1)})$ at $t = \{2, \ldots, T\}$.

Inference of Temporal Model for Customer Preferences

We adapt the collapsed Gibbs sampler used in [10] to our problem setting. The collapsed Gibbs sampler is an effective inference technique for high-dimensional problems, which can integrate out one or more variables such as model parameters, when sampling for some other variables [19]. This allows us to only sample the latent variable for each dimension in turn based on the compound conditional probability distribution. As the product category information is given in $\boldsymbol{Q}$, the product latent variable z_{ij}^I is known in advance. We only need to infer the customer latent variable $z_{ij}^{U(t)}$, which refers to the latent group for customer u_i at t. Given the preference matrix $\boldsymbol{X}^{(t)}$, category l of product, and the latent variables for all customer-product pairs except (u_i, v_j), $\boldsymbol{z}^{\neg(ij)}$ (i.e. $\boldsymbol{z}^{U(t)\neg(ij)}$, we omit $U(t)$ in the formula), the conditional distribution of $z_{ij}^{U(t)}$ is:

$$P(z_{ij}^{U(t)} = k | z^{\neg(ij)}, l, \boldsymbol{X}^{(t)}) \propto \begin{cases} \left(\frac{n_{klr}^{\neg(ij)} + \frac{\beta}{R}}{\sum_r n_{klr}^{\neg(ij)} + \beta}\right)\left(n_{ikt}^{\neg(ij)} + \frac{\alpha^U}{K}\right) & t = 1 \\ \left(\frac{n_{klr}^{\neg(ij)} + \frac{\beta}{R}}{\sum_r n_{klr}^{\neg(ij)} + \beta}\right)\left(n_{ikt}^{\neg(ij)} + \lambda p_{ik}^{(t-1)}\right) & t > 1 \end{cases} \tag{4.1}$$

Specifically, n_{klr} refers to the number of preference level r ($r \in \{1, \ldots, R\}$) for the customer-product joint group (k, l) across all time periods. The superscript $\neg(ij)$ means that the preference x_{ij} of the current customer-product pair (u_i, v_j) is excluded from the counting. Similarly, $n_{ikt}^{\neg(ij)}$ denotes the number of preference records of u_i in customer group k at t, except the one for (u_i, v_j) pair. The prior $\lambda \boldsymbol{p}_i^{(t-1)}$ in the second case of Eq. 4.1 acts as a pseudocount of u_i in the customer group k.

The two terms on the right hand side of Eq. 4.1 shows that the conditional distribution is proportional to the product of two Dirichlet-Multinomial distributions:

(1) the proportion of preference r for customer-product joint group (k, l) among R preference levels;
(2) the proportion of preference records of customer u_i at t falling in customer group k among K customer groups.

In one Gibbs sampling round, we sample $z_{ij}^{U(t)}$ for each customer-product pair (u_i, v_j) in $\{\boldsymbol{X}^{(1)}, \ldots, \boldsymbol{X}^{(T)}\}$, and then we update $\boldsymbol{P}^{(t)}$ ($t \in \{1, \ldots, T\}$) as follows:

$$p_{ik}^{(t)} = \frac{n_{ikt} + \lambda p_{ik}^{(t-1)}}{\sum_{k=1}^{K}(n_{ikt} + \lambda p_{ik}^{(t-1)})} \tag{4.2}$$

where n_{ikt} is counted based on the sampling results in this round. After that, the next sampling round can start and this process iterates until convergence or when it reaches a predefined number of iterations.

After obtaining the final sampling results, the group-level preference $\boldsymbol{B}$ and individual preference $\hat{\boldsymbol{X}}^{(t)}$ can be estimated by:

$$b_{kl} = \sum_{r=1}^{R} r\left(\frac{n_{klr} + \frac{\beta}{R}}{\sum_r n_{klr} + \beta}\right) \tag{4.3}$$

$$\hat{x}_{ij}^{(t)} = \boldsymbol{p}_i^{(t)} \boldsymbol{B} \boldsymbol{q}_j^{(t)} \tag{4.4}$$

The inference procedure of our method is summarised in Table 4.1.

Intuitively, the output $\boldsymbol{p}_i^{(t)} \boldsymbol{B} \in [0, R]^{1 \times L}$ can reflect the preferences of customer u_i for L product categories in any time period t. Therefore, the temporal preferences of N customers can be captured using $\{\boldsymbol{P}^{(1)}, \ldots, \boldsymbol{P}^{(T)}\}$ and $\boldsymbol{B}$.

Table 4.1 The inference procedure of our temporal model for customer preferences

Input	preference matrices $\{\boldsymbol{X}^{(1)}, \ldots, \boldsymbol{X}^{(T)}\}$, product category matrix $\boldsymbol{Q}$
	number of sampling iterations, hyperparameters α^U and β
Output	customer group membership $\{\boldsymbol{P}^{(1)}, \ldots, \boldsymbol{P}^{(T)}\}$
	group-level preference $\boldsymbol{B}$
	for iteration = 1:number_of_sampling_iterations
	for each record (u_i, v_j, t)
	if iteration == 1
	$\alpha^{(t)} = \alpha^U$
	else
	$\alpha^{(t)} = \lambda p_{ik}^{(t-1)}$
	endif
	sample $z_{ij}^{U(t)}$ based on Eq. 4.1
	endfor *% for one record*
	update $\{\boldsymbol{P}^{(1)}, \ldots, \boldsymbol{P}^{(T)}\}$ based on Eq. 4.2
	endfor *% for one sampling iteration*
	estimate $\boldsymbol{B}$ based on Eq. 4.3

4.2.4 Analyse Customer Preference Changes

The third module is to analyse customer preference changes based on $\boldsymbol{p}_i^{(t)}\boldsymbol{B}$ at $t \in \{1, \ldots, T\}$. To further understand if the changes are in the direction targeted by the health program, we label each product category l as either *healthy*, *neutral* or *unhealthy*. Among the 24 food categories in our transaction data:

- 5 categories are labelled as *healthy*: such as vegetables, fruits and packaged salads;
- 5 categories are labelled as *unhealthy*: such as biscuits, soft drinks and confectionery;
- 14 categories are labelled as *neutral*: such as beef, pork and seafood.

Although this manual labelling may be simplistic and coarse-grained, we posit that it generally reflects the accepted health perception of food categories. Our aim is to investigate whether the customer behaviour has changed, and more specifically, whether the customer preferences for the healthy categories have increased due to the participation in the health program.

Based on the category labels, we define a healthy indicator vector $\boldsymbol{h} \in \{0, 0.5, 1\}^L$, and for each category $l \in \{1, \ldots L\}$, the indicator value is:

$$h_l = \begin{cases} 1 & \text{if } l \text{ is } healthy, \\ 0.5 & \text{if } l \text{ is } neutral, \\ 0 & \text{if } l \text{ is } unhealthy. \end{cases} \tag{4.5}$$

Similarly, the indicator vector for unhealthy categories is defined as $\mathbf{1} - \boldsymbol{h}$, which means that the value is 0 for all healthy categories, 1 for all unhealthy categories, and 0.5 for neutral categories.

For customer u_i, we compute the correlation coefficient between the customer preference $\boldsymbol{p}_i^{(t)}\boldsymbol{B}$ and the healthy indicator $\boldsymbol{h}$, and also the correlation coefficient between $\boldsymbol{p}_i^{(t)}\boldsymbol{B}$ and the unhealthy indicator $\mathbf{1} - \boldsymbol{h}$. The difference between these two correlation coefficients defines the *health score* (i.e. hs) for customer u_i at t:

$$hs_i^{(t)} = \rho\big((\boldsymbol{p}_i^{(t)}\boldsymbol{B})^\top, \boldsymbol{h}\big) - \rho\big((\boldsymbol{p}_i^{(t)}\boldsymbol{B})^\top, \mathbf{1} - \boldsymbol{h}\big) \tag{4.6}$$

where $\rho(\boldsymbol{x}, \boldsymbol{y})$ denotes Pearson's correlation coefficient:

$$\rho(\boldsymbol{x}, \boldsymbol{y}) = \frac{E[(\boldsymbol{x} - \mu_x)(\boldsymbol{y} - \mu_y)]}{\sigma_x \sigma_y} \tag{4.7}$$

Given the starting time of the program t_p, we split the T time periods into two phases:

(1) before the program $t \in \{1, \ldots, t_p - 1\}$;
(2) during and after the program $t \in \{t_p, \ldots, T\}$

The change in customer preferences towards the healthy food categories δ_i is quantified by the difference between the average $hs_i^{(t)}$ in the second phase and the average $hs_i^{(t)}$ in the first phase. Formally, the formula of u_i's preference change δ_i is:

$$\delta_i = \underbrace{\frac{1}{T - t_p + 1} \sum_{t=t_p}^{T} hs_i^{(t)}}_{\text{during and after}} - \underbrace{\frac{1}{t_p - 1} \sum_{t=1}^{t_p - 1} hs_i^{(t)}}_{\text{before}} \tag{4.8}$$

The positive value of δ_i indicates an increase in customer u_i's preference towards healthy categories after joining the health program, and a higher value of δ_i implies a greater change in the desirable direction. Therefore, δ_i is our key measure used to evaluate the program impact on customer u_i.

4.2.5 Evaluate Program Impact on Customer Segments

The fourth module is to evaluate the program impact on different customer segments. The values of δ_i for all customers can provide a general understanding of the health program impact. However, we are also interested in determining the types of customers that are more receptive to the health program. This insightful information can inform the design of future personalised programs targeting specific customers.

Generally, the entire customer base can be segmented using a number of features, e.g. geographic, demographic or behavioural [20]. In our study, the self-reported demographic and health information collected through surveys is used for customer segmentation. We consider four features: *gender*, *age*, *who customers live with* (i.e. alone, with partner or with family), and *Body Mass Index* (BMI).

The complete set of customers is partitioned into non-overlapping S_A segments with respect to the value of an attribute A. If the attribute is numeric, e.g. age, its values are discretised into S_A levels. The customers in segment π_s ($s \in \{1, \ldots, S_A\}$) have the same value of A.

As the transaction data of customers who did not participate the health program is unavailable, we split all customers into the *experimental group* and *control group* according to the duration of their participation in the program. Specifically, the experimental group completed two surveys at (1) the start of the program and (2) 12 weeks after the start date, which means that they participated in the entire program; whereas the control group completed only the first survey, so that they participated only in half of the program or less.

Overall, the observed changes δ of individual customers are minor and hard to pick in the short period of one year. Thus, to quantify the impact of the program on customer segment π_s, we sort all the customers in π_s according to their δ_i, and measure the portion of the customers from the experimental group in the set of top-α customers, ranked by δ_i. This reflects whether the customers in the experimental group have greater preference changes towards healthy categories than the customers in the control group.

We define the *impact* of the health program on customer segment π_s as:

$$impact_s = \frac{1}{N_s N_{s@exp}} \sum_{\alpha=1}^{N_s} n_\alpha \tag{4.9}$$

where N_s is the total number of customers in π_s, and $N_{s@exp}$ is the number of customers from π_s who are also in the experimental group. The counter n_α refers to the number of customers in the experimental group and the top-α list. The baseline of $impact_s$ is 0.5. The greater $impact_s$ values imply that the customers in the experimental group are ranked higher than the customers in the control group, which indicates that the health program is more effective for customers in π_s. Importantly, $impact_s$ is computed for each customer segment and is used as the main metric for evaluating the impact of the health program.

4.3 Case Study

This section presents the results of our case study on how the health program influences the purchase behaviour of the program participants. The case study involves 884 customers and 3,394 products from 24 categories, as described in Sect. 4.2.2.

The experimental group comprises 190 customers who participated in the complete program, whereas the remaining 694 customers are in the control group.

Following the procedure described in Sect. 4.2, we convert the transaction data into a customer preference tensor, and then construct the temporal preference model by factorising monthly $\boldsymbol{X}^{(t)}$ into $\boldsymbol{P}^{(t)}$, $\boldsymbol{B}$ and $\boldsymbol{Q}^{\top}$.

The parameters are configured as follows: the number of customer groups $K = 20$, the product groups are the 24 predefined categories ($L = 24$), the number of preference levels $R = 5$, λ is set to 10, and the hyperparameters α^U and β are set to 1 as in [10]. The health program started in May (i.e. $t_p = 5$), so the first phase is $t \in \{1, \dots, 4\}$ and the second phase is $t \in \{5, \dots, 12\}$, when using Eq. 4.8.

4.3.1 *Visualisation of Preference Changes*

We first visualise customer group membership $\boldsymbol{p}_i^{(t)}$ and their preferences for different categories $\boldsymbol{p}_i^{(t)}\boldsymbol{B}$ across 12 months in Figs. 4.3 and 4.4. We select three customers who clearly demonstrate different degrees of variabilities in the group membership and category-level preferences over time.

For each customer, there are two plots: (1) the customer group membership on the left; and (2) the category-level preferences on the right:

(1) For the customer group membership plot, each colour block corresponds to $p_{ik}^{(t)}$, and each column indicates the membership distribution over 20 latent customer groups in one month, with the sum of one column $\sum_k p_{ik}^{(t)} = 1$. Each row tracks how the customer's soft membership of that group changes over time.
(2) For the category-level preference plot, each colour block corresponds to customer u_i's preference for category l in one month. Each column can be used to compare the customer preferences for all 24 categories in one month. Each row reflects the fluctuations of the customer preference for a certain category over time.

Customer 14 (as shown in the first row of Fig. 4.3) had a stable membership over all time periods, consistently belonging to group 12. The customer preferred category 23 (packaged salads) and did not like categories 13 (cheese), 16 (beef) and 17 (lamb), which implies that this customer purchased more vegetables.

Different from customer 14, customer 36 (as shown in the second row of Fig. 4.3) had a mixed membership in groups 7, 10 and 17. It is interesting to note that the membership in group 17 was identified in June, after joining the health program, and it was observed till the end of that year. From the bottom-right plot in Fig. 4.3, the customer's preference for category 20 (vegetables) was consistently high, while her preference for category 22 (fruit desserts) increased gradually, especially after June.

As shown in Fig. 4.4, customer 64 gradually switched from group 20 to group 8 in the middle of the year. The customer's preferences for categories 23 (packaged salads) and 24 (nuts) increased from August, while her preference for category 5 (confectionery) decreased after July.

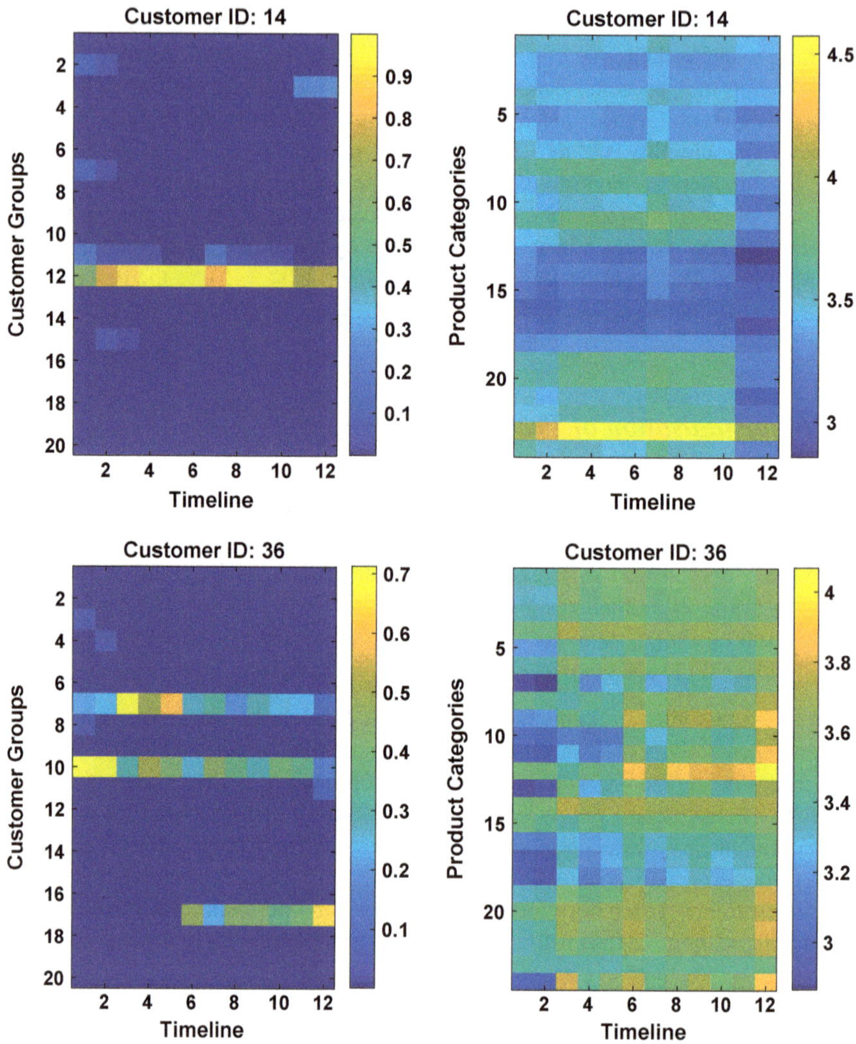

Fig. 4.3 Visualisation of customer group membership (left) and customer preferences (right) for 24 product categories over 12 months (customers 14 and 36) (time unit: month)

The visualisation as shown in Figs. 4.3 and 4.4 can facilitate an intuitive understanding of the temporal preference changes of each individual customer.

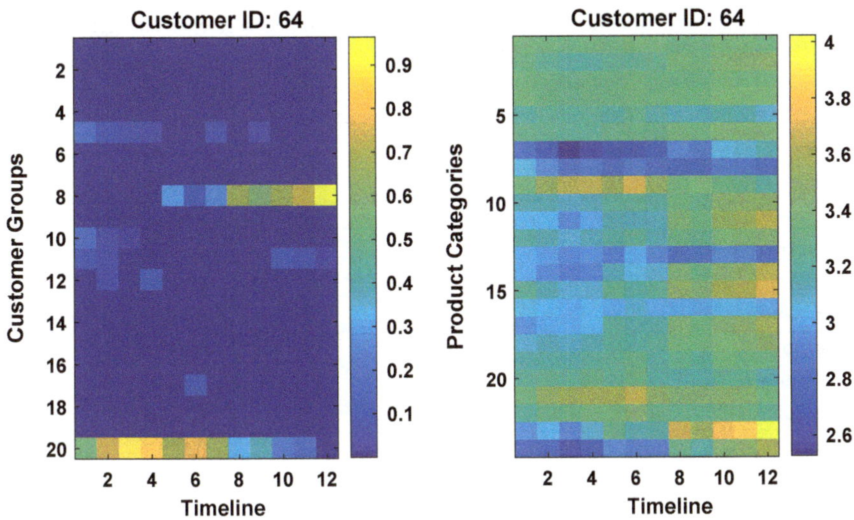

Fig. 4.4 Visualisation of customer group membership (left) and customer preferences (right) for 24 product categories over 12 months (customer 64) (time unit: month)

4.3.2 Program Impact on Different Types of Customers

Besides the visualisation, we also quantify the customer preference changes and compare the impact of our health program on different types of customers.

We first compute the mean preference change (i.e. the mean value of δ_i across all relevant customers) to compare the behaviour changes of the control and experimental group. For the control group, the mean preference change is 0.0204, while for the experimental group, the mean preference change is 0.055, which is more than twice as much as that of the control group. This observation generally shows that the experimental group had a greater preference change towards healthy categories than the control group.

The overall $impact_s$ score of the health program, without segmenting customers, is 0.527. To get an insight of the fine-grained program impact, the customers are partitioned into groups according to four different criteria:

- *gender*: male and female;
- *age*: ≤ 30, (30, 40], (40, 50] and > 50;
- *who customers live with*: live alone, with partner and with family;
- *BMI*: normal, overweight and obese.

For each segment, we report the segment size, the number of customers in the experimental group and the program impact score[1] in Table 4.2.

[1]The customers who had not provided their demographic and health information were excluded from this analysis.

Table 4.2 Impact scores of the health program for different customer segments. The highest impact score for each partition criterion is in bold

Segment	Number of customers		Impact
	Total	Experimental	
Male	349	72	0.4898
Female	517	118	**0.5533**
Age <=30	184	30	**0.5844**
30 < Age <= 40	347	83	0.5431
40 < Age <= 50	232	54	0.5067
Age > 50	100	23	0.4574
Live alone	91	21	0.4929
With partner	344	68	0.5038
With family	385	91	**0.5555**
Normal	329	73	0.5592
Overweight	303	77	0.4787
Obese	220	34	**0.5705**

Segmentation by Gender

As shown in the first section of Table 4.2, the male customers account for about 40% of all customers, and 37.8% of the experimental group. This means that the ratio of male and female customers in the control group and the experimental group are close, both of them being 2:3. The impact score of the health program is 0.5533 for the female customers, which is higher than 0.4898 observed for the male customers. This implies that the female customers in the experimental group were effectively motivated to purchase healthier food, while the male customers in the experimental group were less receptive to the health program. Therefore, *the health program was found to be more effective for female customers*.

Segmentation by Age

The reported age of the customers varied from 19 to 67. The customers are partitioned into four equal-width segments as shown in the second section of Table 4.2. The health program impact score drops significantly from 0.5844 for customers younger than 30 to 0.4574 for customers older than 50. The difference between these two impact scores supports that different customer segments have different responses to the health program. For customers older than 40, the impact scores are 0.5067 and 0.4574, which are close to or lower than the baseline. We notice that the program impact decreases gradually as the age increases. One possible reason is that it might be easier for younger customers to change their dietary habits. However, this does not imply that customers older than 50 purchase unhealthy food. On the contrary, they might purchase healthier food consistently before and after joining the health program, resulting in smaller preference changes between two phases. As our impact measure focuses on the behaviour change, the results show that *the health program did not influence older customers as much as younger customers.*

Segmentation by Who Customers Live With

Based on the survey question "who you live with", the customers are partitioned into three segments: live alone, live with partner and live with family. The segment of customers who live with family is the largest, comprising 46.95% of all customers. The health program impact on this segment is 0.5555, which is higher than the other two segments. For the customers who live alone or with partner, the impact scores are relatively low, being 0.4929 and 0.5038, respectively. Thus, from the perspective of their living arrangement, the results show that *the customers living with family achieved a greater preference change towards healthy foods than the other customers.*

Segmentation by BMI

The BMI is derived from the height and weight of a person,[2] and its value is classified as underweight, normal, overweight and obese [1]. As the proportion of underweight customers is less than 1.5%, we excluded them from this analysis. It is worth noting that the overweight and obese segments together take up 61.4% of all customers, which shows the severity of the overweight problem and the necessity of encouraging people to eat healthily. The obese segment achieves the highest impact 0.5705, closely followed by the normal segment with 0.5592. The overweight segment is left behind, with 0.4787. The impact scores of the obese and normal segments are close, but there is a substantial gap between these segments and the overweight segment. The results suggest that *the health program was more effective for obese and normal weight customers than for overweight customers.*

Discussion

Our results clearly demonstrate that different types of customers were influenced by the health program to a different extent. After grouping customers based on four segmentation criteria, the health program was found to be more effective for female customers, younger customers, customers who live with their family, and obese or normal weight customers.

However, the program impact across all segments was relatively low. There are four main possible reasons:

(1) The offered 10% discount on fruits and vegetables may not be attractive enough for customers to notably change their behaviour.
(2) The lack of real control group may have undermined the significance of the results, since using the customers, who participated in part of the health program as the control group, may not truly represent customers not participating in the health program.

---

[2]The formula is $\mathrm{BMI} = \mathrm{weight(kg)}/\mathrm{height(m)}^2$.

(3) The actual family size is unknown, so that the results may be inaccurate when comparing customers who purchased, for example, for a family of 2 versus for a family of 5.
(4) The one-year duration of the health program may not be sufficiently long to identify stable behaviour changes of the customers.

Despite these shortcomings of the program data, the evaluation results are encouraging and allow the health program stakeholders to get a fine-grained insight into the impact of the health program. In future work, we would like to consider utilising the dataset with real control groups to conduct further evaluation to understand the program impact on different customers.

4.4 Summary

In this chapter, we proposed an approach for tracking the customer preferences over time. We evaluated the impact of our supermarket health program on different types of customer segments, defined by demographic and health attributes such as age, gender, living arrangements and BMI. Based on the purchase data and information collected by the survey of our supermarket health program, we analysed how customers from different segments changed their preferences over time for various food categories. Overall, the results showed that the health program successfully motivated customers to purchase healthier food. The segment-wise impact scores demonstrated that different types of customers were influenced to a different extent. We found that female customers, younger customers, customers who live with their family and obese customers were more receptive to the health program than their counterparts.

Our results can be used to provide guidelines to enhance future health programs and benefit the wider society. The results allow tailoring or personalising programs to motivate customers who were less receptive in this program, such as male customers, senior people and overweight customers. It would be helpful to target them and raise their awareness about the importance of a healthy diet.

Although our method has been designed for a health program, it is a generic method that can be applied to other programs involving transaction records and customer profiles.

References

1. World Health Organization (2015) World health organization factsheet: obesity and overweight. http://www.who.int/mediacentre/factsheets/fs311/en/

2. Ball K, McNaughton SA, Le HND, Gold L, Ni Mhurchu C, Abbott G, Pollard C, Crawford D (2015) Influence of price discounts and skill-building strategies on purchase and consumption of healthy food and beverages: Outcomes of the supermarket healthy eating for life randomized controlled trial. Am J Clin Nutr 101(5):1055–1064
3. Berkovsky S, Hendrie G, Freyne J, Noakes M, Usic K (2015) The HealthierU portal for supporting behaviour change and diet programs. In: Proceedings of 23rd Australian national health informatics conference 2015, vol 214. IOS Press, pp 15–21
4. Chen M-C, Chiu A-L, Chang H-H (2005) Mining changes in customer behavior in retail marketing. Expert Syst Appl 28(4):773–781
5. Liu B, Hsu W, Han H-S, Xia Y (2000) Mining changes for real-life applications. In: Proceedings of the 2nd international conference on data warehousing and knowledge discovery, 2000, vol 1874. Springer, pp 337–346
6. Masseglia F, Poncelet P, Teisseire M (2003) Incremental mining of sequential patterns in large databases. Data Knowl Eng 46(1):97–121
7. Koren Y (2010) Collaborative filtering with temporal dynamics. Commun ACM 53(4):89–97
8. Xiong L, Chen X, Huang T-K, Schneider JG, Carbonell JG (2010) Temporal collaborative filtering with Bayesian probabilistic tensor factorization. In: Proceedings of the 2010 SIAM international conference on data mining, vol 10. SIAM, pp 211–222
9. Fanaee-T H, Gama J (2015) Multi-aspect-streaming tensor analysis. Knowl Based Syst 89:332–345
10. Li B, Zhu X, Li R, Zhang C, Xue X, Wu X (2011) Cross-domain collaborative filtering over time. In: Proceedings of the 22nd international joint conference on artificial intelligence, 2011. AAAI Press, pp 2293–2298
11. Ekstrand MD, Riedl JT, Konstan JA (2011) Collaborative filtering recommender systems. Found Trend Hum Comput Interact 4(2):81–173
12. Billsus D, Pazzani MJ (1998) Learning collaborative information filters. AAAI workshop on recommender systems 98:46–54
13. Hofmann T (1999) Probabilistic latent semantic indexing. In: Proceedings of the 22nd international ACM SIGIR conference on research and development in information retrieval, 1999. ACM, pp 50–57
14. Hofmann T (2004) Latent semantic models for collaborative filtering. ACM Trans Inf Syst 22(1):89–115
15. Porteous I, Bart E, Welling M (2008) Multi-HDP: a non parametric Bayesian model for tensor factorization. In: Proceedings of the 23rd AAAI conference on artificial intelligence, 2008, vol 8, pp 1487–1490
16. Geman S, Geman D (1984) Stochastic relaxation, Gibbs distributions, and the Bayesian restoration of images. IEEE Trans Pattern Anal Mach Intell 6:721–741
17. Andrieu C, De Freitas N, Doucet A, Jordan MI (2003) An introduction to MCMC for machine learning. Mach Learn 50(1–2):5–43
18. Bishop CM (2006) Pattern recognition and machine learning. Springer
19. Liu JS (1994) The collapsed Gibbs sampler in Bayesian computations with applications to a gene regulation problem. J Am Stat Assoc 89(427):958–966
20. Baker GA, Burnham TA (2001) Consumer response to genetically modified foods: market segment analysis and implications for producers and policy makers. J Agric Resour Econ 387–403

Chapter 5
Discovering Purchase Behaviour Patterns

5.1 Background

Various factors could impact purchase time, ranging from personal necessity, preferences and seasonal effects, to marketing variables such as products, price and promotions. In this chapter, we build temporal model to uncover both the customer's long-term purchase patterns, which may be caused by gradual preference changes, and also the customer's short-term purchase patterns, which may be caused by regular promotions. The discovered patterns could provide insights of customer behaviour and distinguish between groups of customers in terms of their responses to promotions.

Through exploring the purchase records of some popular products, we notice that the daily sales quantity of a product varied through the whole year. Take the 2-L Coca-Cola as an example, the number of bottles sold each day and the daily price are shown in Fig. 5.1. The price curve shows that there were 13 promotions for this product in the whole year, about once every two weeks. The number of bottles sold each day increased when the price decreased, which is reflected in the 13 peaks corresponding to the promotions. In addition, the quantity sold got a high value before the end of the year around Christmas. Overall, the daily sales quantity is negatively correlated with the price (Pearson's correlation $= -0.4842$). Based on above observations, we assume that the purchase behaviour of an individual customer does not remain the same over time, and it is affected by contextual factors such as weather and promotions.

The stakeholders often desire to know what customers are receptive to their promotions. Therefore, besides the aggregated behaviour of all customers, it is also important to understand the differences among individual customers. However, for a certain product, the individual purchase events may be too sparse to analyse customer behaviour patterns and determine whether they have been affected by promotions. Therefore, we propose to identify shared behaviour patterns with mixture models for customer behaviour analysis.

Problem Formulation and Organisation

Given the customer purchase data, our research task is to discover the latent types of behaviour patterns through a soft-membership customer segmentation and explore

L. Luo, *Temporal Modelling of Customer Behaviour*,
Springer Theses, https://doi.org/10.1007/978-3-030-18289-2_5

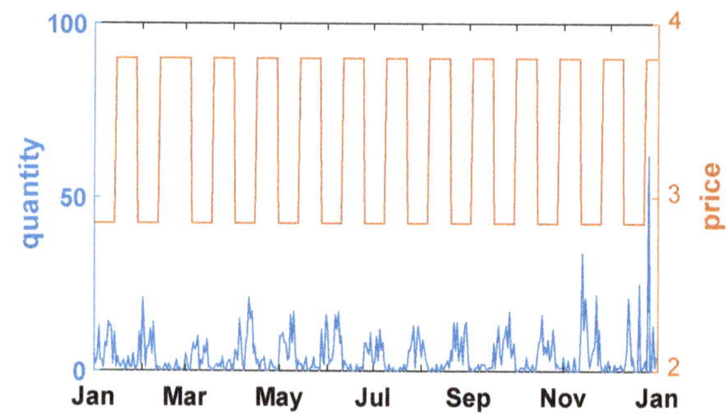

Fig. 5.1 Daily price (orange curve) and number of bottles sold (blue curve) for 2-L Coca-Cola are negatively correlated

how different customers respond to promotions. We consider typical temporal behaviour patterns, including long-term patterns which show the gradual preference drifts over several months and short-term patterns such as the fortnightly regular shopping or the purchase motivated by promotions. An individual customer behaviour can thus be a combination of different types of patterns. For example, although the overall quantity of a product sold exhibits a negative correlation with its price, there might exist two latent customer groups—one with a higher purchase rate during a promotion and another one with stable purchase rate regardless of the promotion. Based on the soft membership of these two groups, the customers who have a stronger association with the first group are considered to be receptive to promotions.

We propose a mixture of Non-Homogeneous Poisson Processes (NHPP) to discover latent customer groups and conduct the soft-membership customer segmentation based on the dynamically observed purchase behaviour. The purchase behaviour pattern of each group is modelled by the intensity function of NHPP, which integrates:

(1) a polynomial component for the long-term patterns due to preference drifts or seasonal effects;
(2) a periodic component for the short-term patterns driven by customer needs and promotions.

We adopt the Expectation Maximisation (EM) algorithm to find the latent groups as well as estimate the coefficients of the intensity functions.

This chapter aims to model the long-term and short-term behaviour patterns explicitly by the intensity function of NHPP. In Chap. 4, we only estimate and track the preference levels at different time periods, without identifying specific customer behaviour patterns. This chapter is organised as follows. Section 5.2 describes the preliminaries, including Poisson processes for purchase behaviour modelling, mixture models and how to apply the EM algorithm to infer the parameters of mixture models. Then Sect. 5.2 continues to explain three modules of our methodology in detail. Section 5.3 introduces our case study on the supermarket purchase data and the behaviour patterns detected by the proposed method, and discusses customers' responses to promotions. Section 5.4 summarises our findings and contributions of this chapter.

Contributions

We conduct a case study on the supermarket purchase data, and our contributions are:

- We validate that the NHPP model can estimate the purchase events accurately, significantly outperforming the baseline method.
- We discover the representative long-term and short-term purchase behaviour patterns of customer groups, which captures the dynamics in purchase behaviour and helps to understand different customers and products.
- We explore the impact of promotions on different customers and find the customers who are receptive to price reductions.

Our proposed approach to discovering temporal patterns of purchase behaviour with different responses to promotions can act as a backbone of a more effective target marketing.

Related Work

Extensive research on customer behaviour analysis has been carried out to provide decision support for business management. Among different approaches, researchers have dedicated considerable efforts to using stochastic models to characterise customer behaviour. Following the classical negative binomial model proposed in [1], several extensions [2–4] focused on modelling the number of visits or products sold over the whole observation period of all customers. The weakness of these models is that they only estimated the overall number of purchase events, without supporting customer segmentation based on the heterogeneity of customers.

To consider the heterogeneity of customers, previous work used demographic information or purchase behavioural variables (e.g. usage rates, loyalty status, and response to a product) to segment customers into groups. For instance, Bucklin et al. [5, 6] proposed the mixture of logit models to segment customers based on their purchase time, purchase quantity and brand choice, but their work did not consider the dynamics of purchase behaviour.

Recent models took behaviour changes into account while conducting customer segmentation, as it allows stakeholders to evaluate their policies and marketing campaigns on different customers [7–10]. Iwata et al. [11] adapted topic models to track the changes of customer interests and item trends. However, their model could not provide specific behaviour patterns regarding the marketing variables for each latent group, making the results difficult to interpret. The hierarchical time-rescaling model proposed in [12] could identify various temporal purchase patterns via point processes. However, this work only modelled at the individual level, which could not deal with sparse purchase records.

To the best of our knowledge, there are no reported methods that could identify interpretable long-term and short-term patterns of purchase behaviour from sparse purchase records. Additionally, our model can analyse the responses of customers to the dynamic marketing variables like sales promotions based on their behaviour patterns.

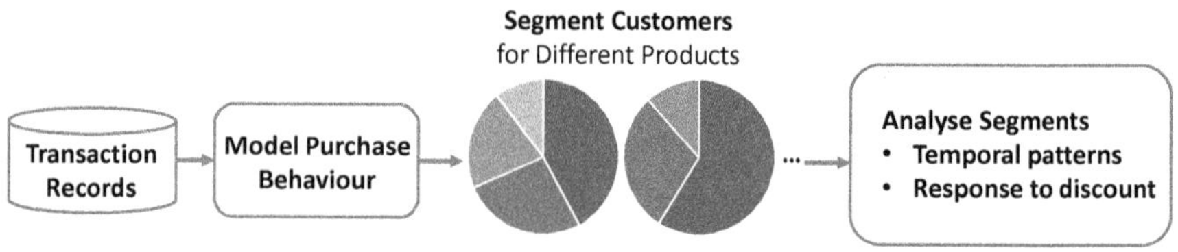

Fig. 5.2 Flowchart of our model

5.2 Methodology

In this section, we introduce our approach for discovering latent customer segments[1] with different long-term and short-term purchase behaviour patterns. We outline the main modules of our method in Fig. 5.2:

(1) *Model purchase behaviour*: we build a mixture of NHPP models for customer purchase behaviour based on the transaction records.
(2) *Segment customers*: for different products: we adopt the EM algorithm to infer the group membership of each customer and estimate the coefficients of the NHPP intensity functions.
(3) *Analyse segments*: we conduct a comprehensive analysis to reveal long-term and short-term temporal patterns of the non-stationary purchase behaviour and provide insights of how customers from different segments respond to promotions.

5.2.1 Preliminaries

Poisson Process for Purchase Behaviour

The Poisson process is a simple yet powerful stochastic process [13] that describes the number of random points in a temporal or spatial space, such as the occurrence of incidents at an intersection, arrivals of customers to a service centre, and activities in social media [14]. The Poisson process can be equivalently represented by the counting process $\{N(t)\}$ $(t \geq 0)$, where $N(t)$ is a random variable for the number of arrivals in the interval $[0, t]$. The counting variable $N(t)$ has a Poisson distribution with parameter Λ, so that the probability density function and expectation of $N(t)$ are:

$$P(N(t) = n) = e^{-\Lambda}\frac{\Lambda^n}{n!} \qquad (5.1)$$
$$E(N(t)) = \Lambda$$

[1]The term 'segment' and 'group' are used interchangeably in this chapter.

If the intensity λ of the Poisson process is a constant, it is called a *Homogeneous* Poisson Process (HomoPP), and $\Lambda = \lambda t$. However, if λ varies over time, which means that the corresponding counting process has non-stationary increments, the Poisson process is a *Non-Homogeneous* Poisson Process (NHPP). The varying intensity can be described by a function $\lambda(x)$, and $\Lambda = \int_0^t \lambda(x)dx$.

In terms of the customer purchase behaviour analysis, the purchase events for a product could be described by a Poisson process, so that the number of purchase events up to t is a counting process $\{N(t)\}$. We provide more details about the purchase model in Sect. 5.2.2.

Mixture Models

The mixture model [15] is the fundamental model of our customer segmentation. Given observations distributed according to a mixture of K components, the probability of observing x is:

$$p(x) = \sum_{k=1}^{K} \phi_k p_k(x) \tag{5.2}$$

where $\phi_k \in [0, 1]$ is the weight of component k in explaining the observation x, and the distribution of component k is $p_k(x)$. The weights of all components satisfy the condition that $\sum_{k=1}^{K} \phi_k = 1$. If the distribution of each component can be described by parameter $\mathbf{\Theta}$, we have $p_k(x) := p(x|\mathbf{\Theta})$.

For example, the widely used mixture model—Gaussian Mixture Model (GMM)—assumes that the observations are generated from a mixture of a finite number of Gaussian distributions. The probability distribution of x is a linear combination of K Gaussian distributions as follows:

$$p(x) = \sum_{k=1}^{K} \phi_k p(x|\boldsymbol{\mu}_k, \mathbf{\Sigma}_k) \tag{5.3}$$

where $\boldsymbol{\mu}_k$ and $\mathbf{\Sigma}_k$ are the mean and covariance of component k, respectively. GMM provides more flexibility in describing observations than using a single Gaussian distribution.

From another viewpoint, assuming each observation x is associated with an unobserved discrete random variable z with K possible values $\{1, \ldots, K\}$, the marginal distribution of x can be represented by summing the joint distribution $p(x, z)$ over all possible z, and the equation is as follows:

$$\begin{aligned} p(x) &= \sum_{k=1}^{K} p(x, z = k) \\ &= \sum_{k=1}^{K} p(x|z = k)p(z = k) \end{aligned} \tag{5.4}$$

where $p(x|z = k)$ is the likelihood of x when the latent variable is k, which corresponds to $p_k(x)$ in Eq. 5.2; and $p(z = k)$ is the prior of the latent variable z, which corresponds to the weight of component k, ϕ_k, in Eq. 5.2.

Based on the Bayes' rule, given the observation x, the posterior probability of $z = k$ is:

$$\begin{aligned} p(z = k|x) &= \frac{p(x, z = k)}{p(x)} \\ &= \frac{p(x|z = k)p(z = k)}{\sum_{s=1}^{K} p(x|z = s)p(z = s)} \end{aligned} \tag{5.5}$$

The denominator is the sum of joint distributions over all possible z, so we have $\sum_{k=1}^{K} p(z = k|x) = 1$. If we consider the mixture model as a generalisation of K-means clustering with K groups (i.e. components), the posterior probability $p(z = k|x)$ is the soft membership of x in group k, which means the responsibility of group k in explaining x [15].

In summary, for the mixture models, we should infer the following parameters:

- parameter sets for all groups $\mathbf{\Theta}$ (i.e. likelihood $p(x|z = k)$);
- weights of all groups $\mathbf{\Phi}$ (i.e. prior $p(z)$);
- soft membership $\mathbf{\Pi}$ of all observations (i.e. posterior probability $p(z = k|x)$).

Given $\boldsymbol{x} = \{x_1, \ldots, x_N\}$ with latent variables $\boldsymbol{z} = \{z_1, \ldots, z_N\}$, the log likelihood of the observations can be written as:

$$\begin{aligned} \ell(\mathbf{\Theta}, \mathbf{\Phi}) &= \sum_{i=1}^{N} \log p(x_i; \mathbf{\Theta}, \mathbf{\Phi}) \\ &= \sum_{i=1}^{N} \log \sum_{k=1}^{K} p(x_i|z_i = k; \mathbf{\Theta}) p(z_i = k; \mathbf{\Phi}) \end{aligned} \tag{5.6}$$

Due to the summation inside the log function, it is difficult to obtain the closed-form solution by maximising $\ell(\mathbf{\Theta}, \mathbf{\Phi})$ directly. The Expectation-Maximisation (EM) algorithm [16] is an effective tool to solve this problem, and we will briefly introduce EM in the following section.

Expectation-Maximisation Algorithm

The EM algorithm is often applied to the maximum likelihood estimation with latent variables, such as Eq. 5.6. Since it is difficult to maximise the log likelihood ℓ directly, EM optimises the parameters in an iterative manner via E- and M-steps.

Generally, in the E-step, the algorithm constructs a lower-bound on ℓ [17], which is defined as:

$$\ell(\mathbf{\Theta}, \mathbf{\Phi}) \geq \sum_{i=1}^{N} \sum_{k=1}^{K} Q_i(z_i = k) \log \frac{p(x_i|z_i = k; \mathbf{\Theta}) p(z_i = k; \mathbf{\Phi})}{Q_i(z_i = k)} \tag{5.7}$$

Equation 5.7 moves the summation out of the log function according to the Jensen's inequality. Moreover, $Q_i(z_i = k)$ can use any distributions, but to make sure the lower-bound is tight for current $\mathbf{\Theta}$, it is often set as the posterior distribution $p(z_i = k|x_i)$. In the M-step, the algorithm maximises the lower-bound by adjusting $\mathbf{\Theta}$ and $\mathbf{\Phi}$. The E- and M-steps are conducted iteratively until convergence.

Specifically, for mixture models introduced above, the workflow to infer $\mathbf{\Theta}$, $\mathbf{\Phi}$ and $\mathbf{\Pi}$ is as follows:

(1) Initialise $\mathbf{\Theta}$ and $\mathbf{\Phi}$.
(2) Iterate between E- and M-steps until convergence:

- E-step: based on initial $\mathbf{\Theta}$ and $\mathbf{\Phi}$ or their values from the previous iteration, update $\mathbf{\Pi}$.
- M-step: based on $\mathbf{\Pi}$ from E-step, update $\mathbf{\Theta}$ and $\mathbf{\Phi}$.

It is worth noting that the EM algorithm may be slow to converge and sensitive to the initialisation. For example, for GMM, the K-means clustering can be run in advance to provide a suitable initialisation for EM [15]. Besides mixture models, the EM algorithm can also be applied to other scenarios such as when the unobserved variables are random missing data.

5.2.2 Model Purchase Behaviour via NHPP

As discussed in Sect. 5.1, the increment of $\{N(t)\}$ is indeed an NHPP, which could be affected by various contextual factors. In this section, we would like to explore the insights of purchase behaviour through the intensity function $\lambda(x)$ of the counting process of purchase events. We use $\lambda(x)$ to capture the temporal patterns over long-term and short-term observations.

The *long-term patterns* can be influenced by factors such as customer preference changes and seasonal effects (e.g. weather and seasonal price). We select the polynomial function to model gradual long-term changes, considering the trade-off between tractability and flexibility of the model. As the main challenge in the estimation procedure is the integral $\int_0^t \lambda(x)dx$ [12], using the polynomial function could generate the closed-form solution efficiently, and it is capable of capturing the typical long-term dynamics of purchase behaviour.

The *short-term patterns* are used to describe periodic (e.g. weekly, fortnightly or monthly) purchase events, which are mainly driven by the customer needs or attractiveness of promotions. For example, if the customers' favourite soft drinks are on sale, they may make a temporary decision to purchase the product and stockpile some bottles to use later. In this case, the purchase events are driven by the promotion. In the context of supermarket, the personal needs and the promotion of a product are generally periodic, so we use a sine function to model the short-term pattern of the purchase behaviour.

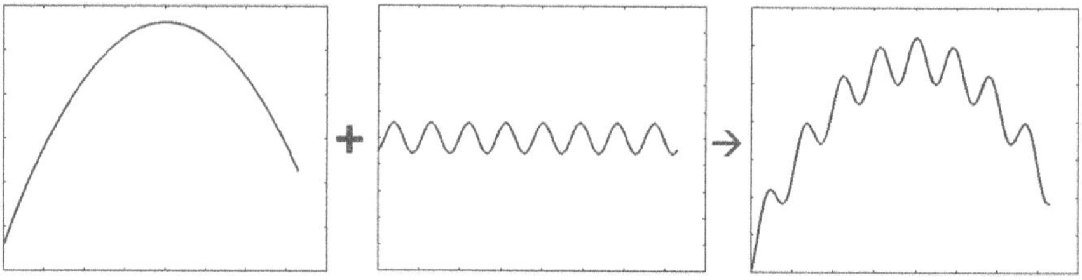

Fig. 5.3 Integrate the polynomial component (left) and the periodic component (middle) to get the purchase rate curve (right)

Therefore, the intensity function can be an integration of a *polynomial component* and a *periodic component* as illustrated in Fig. 5.3.

Formally, we define the intensity function $\lambda(x)$ as:

$$\lambda(x) = \sum_{d=0}^{D} w_d x^d + a \sin(bx + c) \tag{5.8}$$

with the restriction that $\lambda(x) \geq 0$ for any x. The term $\lambda(x)$ acts like a density function for purchase events, so higher $\lambda(x)$ values correspond to more frequent purchase events, and vice versa. The polynomial component fits the trend of the purchase rate, including typical patterns such as increase, decrease, U-shape or inverse U-shape, depending on w_d. The parameter D is the degree of the polynomial component, which is tuned to the dataset (usually $D = 2$ suffices). For the sine component, a, b, c are the amplitude, frequency and phase of the short-term patterns, respectively. If $w_d = 0$ for $d > 0$ (i.e. the polynomial component is a constant w_0), the purchase rate only has periodic patterns; if $a = 0$, the purchase rate only has long-term trends; if both w_d and a are zeros, the purchase behaviour is a HomoPP with $\lambda = w_0$.

Estimate Purchase Intensity

We are interested in the daily purchase behaviour, so we set the granularity of our analysis to 1 day and ignore the fine-grained details such as hours or minutes in the transaction records. If $N(t)$ denotes the number of purchase events up to day t, the daily purchase event can be computed based on the following equation:

$$\Delta N(t) = N(t) - N(t-1),\ t \in \{1, \dots, T\} \tag{5.9}$$

When $\Delta N(t)$ ($\Delta N(t) \geq 0$) is non-zero, it means that there is at least one purchase event occurring at t; otherwise, the customer did not buy that product at t. The expectation of $\Delta N(t + \delta)$ for a small time interval δ is:

$$\begin{aligned} E(\Delta N(t+\delta)) &= E(N(t+\delta)) - E(N(t)) \\ &= \Lambda(t+\delta) - \Lambda(t) \\ &= \int_t^{t+\delta} \lambda(x) dx \end{aligned} \tag{5.10}$$

For a small δ, the value of $\lambda(t+\delta)$ represents the instantaneous purchase rate at $t+\delta$. Assuming the intensity is constant during $(t, t+\delta]$, the unbiased estimation of $\lambda(t+\delta)$ is $E(\Delta N(t+\delta))/\delta$ according to the parameter estimation for HomoPP [13]. As the granularity level of our model is 1 day, the underlying non-homogeneous Poisson intensity can be viewed as a piece-wise step function with constant purchase rate within each day. Thus, we can estimate coefficient θ (includes w_d, a, b, c in Eq. 5.8) of $\lambda(x)$ by maximising the likelihood of the daily number of purchase events $\Delta N(t), t \in \{1, \ldots, T\}$.

5.2.3 Segment Customers with Mixture Models

To understand the purchase behaviour of an individual customer, one option would be to fit $\lambda(x)$ to their purchase events of a certain product. However, the purchase events of an individual may be too sparse for an analysis of behaviour patterns and responses to promotions. Moreover, the stakeholders often desire to get higher-level information such as:

(1) What are the different types of behaviour patterns?
(2) What is the proportion of customers with a certain type of behaviour?

Therefore, a mixture of NHPP is utilised to segment customers based on the purchase behaviour patterns characterised by the intensity function in Eq. 5.8. Given a transaction dataset with U customers and M products, for a target product m, our tasks are to:

(1) Extract all customers $\{u_i\}^{(m)}$, $i \in \{1, 2, \ldots, U\}$, who bought m;
(2) Identify K latent segments of $\{u_i\}^{(m)}$, based on customer u_i's $N_i(T)^{(m)}$ purchase events during the observation period $[0, T]$, where $N_i(T)^{(m)}$ is the total number of purchases of m by u_i;
(3) Estimate u_i's soft membership in K latent segments and the size of each customer segment.

As the following description is all in the context of the purchase behaviour of a certain product m, we omit the superscript m.

Each component in the mixture of NHPP corresponds to a latent customer segment, and the purchase behaviour of all customers in segment k shares the same $\lambda_k(x)$. The behaviour of an individual customer is a combination of $\lambda_k(x)$ of multiple segments, weighted by u_i's soft membership π_{ik} in segment k, $\sum_{k=1}^{K} \pi_{ik} = 1$. In order to discover K latent segments, we estimate the following parameters:

- $\mathbf{\Theta}$: coefficients of $\lambda_k(x)$ for K segments. The coefficient for segment k is θ_k, which includes $\{w_{dk}, a_k, b_k, c_k\}$.
- $\mathbf{\Phi}$ is a $K \times 1$ vector for the proportions of K segments, where $\phi_k \in [0, 1]$ is the proportion of mixture component k, and $\sum_{k=1}^{K} \phi_k = 1$.
- $\mathbf{\Pi}$ is a $U \times K$ matrix of soft membership over K segments for all customers, where $\pi_{ik} \in [0, 1]$ and $\sum_{k=1}^{K} \pi_{ik} = 1$.

Given the time stamps of u_i's purchase events $\boldsymbol{x}_i \in (0, T]$, the element x_{ij} ($j \in \{1, \ldots, N_i(T)\}$) is the time of the jth purchase of u_i. The log likelihood of u_i's purchase events can be written as:

$$\ell(\boldsymbol{\Theta}, \boldsymbol{\Phi}) = \sum_{i=1}^{U} \log \sum_{k=1}^{K} p(\boldsymbol{x}_i | z_i = k; \boldsymbol{\Theta}) p(z_i = k; \boldsymbol{\Phi}) \tag{5.11}$$

where $z_i \in \{1, \ldots, K\}$ is the index of the latent segment for u_i, and $\boldsymbol{x}_i$ has been generated by NHPP with the intensity function $\lambda_k(x)$. As explained in Sect. 5.2.1, it is hard to maximise Eq. 5.11 directly, so we use the EM algorithm to construct a mixture of NHPP for the observations and infer the customer membership and coefficients for $\lambda_k(x)$ iteratively.

The input of the algorithm includes the number of segments K, the parametric form of $\lambda_k(x)$, and the purchase records $\{\boldsymbol{x}_i\}, i \in \{1, 2, .., U\}$. Then the algorithm starts from the E-step and iterates between the E- and M-steps.

E-step

If it is the first iteration, we assign u_i to a segment randomly or based on a predefined initialisation rule. From the second iteration, we use the estimation of $\boldsymbol{\Theta}$ and $\boldsymbol{\Phi}$ from the M-step of the previous iteration to infer new $\boldsymbol{\Pi}$. The posterior probability of u_i in segment k is:

$$p(z_i = k | \boldsymbol{x}_i, \boldsymbol{\Theta}, \boldsymbol{\Phi}) = \frac{p(\boldsymbol{x}_i | z_i = k, \boldsymbol{\Theta}) p(z_i = k, \boldsymbol{\Phi})}{\sum_{s=1}^{K} p(\boldsymbol{x}_i | z_i = s, \boldsymbol{\Theta}) p(z_i = s, \boldsymbol{\Phi})} \tag{5.12}$$

For a Poisson process, given n independent and ordered ($0 < x_{i1} < \ldots < x_{in} \leq T$) observations at $\{x_{ij}\}$ ($j \in \{1, \ldots, n\}$), the likelihood of these observations is:

$$\begin{aligned} p(\boldsymbol{x}_i | z_i = k, \boldsymbol{\Theta}) &= n!\ p(\{x_{ij}\} | n, z_i, \boldsymbol{\Theta}) p(n, z_i, \boldsymbol{\Theta}) \\ &= n! \Big(\prod_{j=1}^{n} \frac{\lambda_k(x_{ij})}{\Lambda_k} \Big) e^{-\Lambda_k} \frac{\Lambda_k^n}{n!} \\ &= \Big(\prod_{j=1}^{n} \lambda_k(x_{ij}) \Big) e^{-\int_0^T \lambda_k(x) dx} \end{aligned} \tag{5.13}$$

In the second line of Eq. 5.13, the term $\lambda_k(x_{ij})/\Lambda_k$ is the probability of observing an event at x_{ij}, so the joint probability of n independent and identically distributed (i.i.d.) observations at $\{x_{ij}\}$ is $\prod_{j=1}^{n}(\lambda_k(x_{ij})/\Lambda_k)$. As there are $n!$ permutations to sample n i.i.d. time points that could result in the order $0 < x_{i1} < x_{i2} < \ldots < x_{in}$, the probability of the ordered observations is $n! \prod_{j=1}^{n}(\lambda_k(x_{ij})/\Lambda_k)$ [13]. Finally, the last term $e^{-\Lambda_k}(\Lambda_k^n/n!)$ is the probability of observing n events in a Poisson process.

Particularly, the posterior probability $p(z_i = k | \boldsymbol{x}_i, \boldsymbol{\Theta}, \boldsymbol{\Phi})$ in Eq. 5.12 can be denoted by π_{ik} and interpreted as u_i's soft membership in segment k. After the segment membership for all customers being successfully updated for this iteration, the M-step is ready to start.

M-step

In the M-step, we estimate the values of $\mathbf{\Theta}$ and $\mathbf{\Phi}$ based on the sequence of daily purchase event $\{\Delta N(t)\}$ $(t \in \{1, \ldots, T\})$, and $\mathbf{\Pi}$ obtained in the E-step.

The segment-level daily purchase event $\{\Delta N_k(t)\}$ $(t \in \{1, \ldots, T\})$ aggregates the purchase events of all customers in the segment. It considers the purchase event increment $\Delta N_i(t)$ of u_i, and u_i's soft membership π_{ik} in k, which is defined as:

$$\Delta N_k(t) = \sum_{i=1}^{U} \pi_{ik} \Delta N_i(t) \tag{5.14}$$

The sequence of $\{\Delta N_k(t)\}$ is used to estimate θ_k by maximising the log likelihood:

$$\hat{\theta}_k = \underset{\theta_k}{argmax} \sum_{t=1}^{T} \log p(\{\Delta N_k(t)\}|\theta_k) \tag{5.15}$$

Finally, we update the size of segment k using the average soft membership in segment k across all customers:

$$\phi_k = \frac{\sum_{i=1}^{U} \pi_{ik}}{U} \tag{5.16}$$

Table 5.1 The inference procedure of mixture of NHPP

Input	number of segments: K, parametric form of $\lambda_k(x)$
	purchase records: $\boldsymbol{X}^{U \times M}$
Output	coefficients of $\lambda_k(x)$: $\mathbf{\Theta}$, size of segments: $\mathbf{\Phi}$
	soft membership: $\mathbf{\Pi}$
	initialise $\mathbf{\Theta}$, $\mathbf{\Phi}$
	while convergence_condition == false
	% **E-step**
	$\mathbf{\Theta}^{old} \leftarrow \mathbf{\Theta}$
	$\mathbf{\Phi}^{old} \leftarrow \mathbf{\Phi}$
	calculate $\mathbf{\Pi}$ based on $p(z_i = k|\boldsymbol{x}_i, \mathbf{\Theta}^{old}, \mathbf{\Phi}^{old})$ (Eq. 5.12)
	% **M-step**
	calculate $\mathbf{\Theta}$ based on (Eqs. 5.14 and 5.15)
	calculate $\mathbf{\Phi}$ based on (Eq. 5.16)
	endwhile

The algorithm iterates until convergence. The convergence criterion is that the soft membership of customers remain stable between consecutive iterations. The final outputs are the estimations of $\mathbf{\Phi}$, $\mathbf{\Pi}$ and $\mathbf{\Theta}$ in the last iteration. The inference procedure is summarised in Table 5.1.

5.2.4 *Analyse Customer Segments*

Based on the outputs of mixture NHPP models and soft customer segmentation result, we perform a comprehensive purchase behaviour analysis and compare segments with respect to the following features:

(1) segment size ϕ_k;
(2) total number of purchase events of a segment $\int_0^T \lambda_k(x)dx$;
(3) long-term patterns such as increase or decrease of purchase rate, determined by w_d of the polynomial component;
(4) cycle length of the periodic short-term patterns, computed by $2\pi/b$ (b is the frequency of the sine component).

Different temporal patterns can reflect the heterogeneity of purchase behaviour between customer segments, even when two segments have similar total numbers of purchase events.

More importantly, we are interested to discover whether the purchase rate of a customer segment is influenced by promotions. We compute the correlation r_k between $\lambda_k(x)$ and the price curve. This idea is similar to checking the price elasticity of demand, which is computed by the percentage of demand change divided by the percentage of price change. For a fixed change rate, e.g. 1 or 10%, the elasticity measures the responsiveness of demand to the price change. The correlation is more appropriate to our goal, as it could reflect the relationship between the demand and price over the whole observation period, rather than a certain change rate as in the price elasticity metric. If r_k is negative, segment k is receptive to promotions, which means that customers purchase more frequently when the price is lower. Otherwise, segment k is not receptive to promotions, which means that the customers would keep their purchase patterns regardless of price reductions.

5.3 Case Study

In this section, we first describe how to prepare the transaction data for the case study and configure the parameters of the model. Then, we concentrate on the following tasks:

(1) Evaluating the capability of the proposed method for modelling the purchase events of individual customers;
(2) Discovering long-term and short-term purchase patterns of latent customer segments and understanding the differences among segments;
(3) Identifying customer segments that are receptive to product promotions.

5.3.1 Data Preprocessing

We notice that supermarket promotions generally involve a series of products from the same brand. For example, biscuits from the same brand with different flavours are included in one promotion. Customers would choose from these products and buy one or a mixed collection of flavours in one transaction. As the ultimate goal is to understand how customers respond to promotions, we choose "brand" as the granularity level of our analysis and aggregate the purchase records of the same brand for customer purchase behaviour modelling.

We select a number of top brands based on the popularity—the number of days when they were bought by customers, from representative food categories. In total, 27 brands from 7 categories are used in this study:

- 3 brands from Ice Cream;
- 3 brands from Soft Drinks;
- 3 brands from Biscuits;
- 4 brands from Snacks;
- 5 brands from Confectionery;
- 5 brands from Chilled Desserts (cream and yoghurt);
- 4 brands from Cereal.

The names of all selected brands are listed in Table 5.2. For each brand, the active customers are chosen based on whether they bought products from that brand more than 10 times in one year. The model construction and customer segmentation are based on the representative brands and active customers.

To explore customer responses to promotions, we need the price of each brand. As there are no actual prices available in the data, we take three steps as follows to compute the *brand price*:

(1) Extract the product price: We infer the daily product price from the purchase quantity and the cost on each day. If there are no purchase records on a certain day, the valid price from the preceding days is used or the most frequent price is used when the closest valid price is more than 1 week old.
(2) Normalise the daily product price to (0, 1], using the ratio of the daily price to the maximal price of the product. This step can eliminate the differences of magnitude among prices of various products, but keep the daily discount rate. As our goal is to explore the customer responses to promotions, the fluctuations of normalised price are more important than the absolute prices.

Table 5.2 The size and estimated number of purchase events of each segment

ID	Category	Brand name	Size of segment				Estimated purchase events			
			1	2	3	4	1	2	3	4
1	Ice cream	Own brand	0.56	0.28	0.16		15.79	13.77	29.31	
2		Peters	0.72	0.21	0.07		14.08	12.34	32.11	
3		Streets	0.77	0.23			12.68	30.77		
4	Soft drinks	Coca-Cola	0.29	0.56	0.08	0.07	49.87	27.56	85.00	163.14
5		Own brand	0.37	0.25	0.33	0.05	26.75	34.54	51.94	98.78
6		Schweppes	0.48	0.15	0.27	0.10	13.61	25.64	14.83	46.13
7	Biscuits	Arnott's	0.51	0.30	0.17	0.02	25.63	36.69	51.84	90.16
8		Own brand	0.30	0.50	0.18	0.02	19.20	19.38	33.51	70.19
9		Fantastic	0.45	0.42	0.13		12.09	15.77	29.44	
10	Snacks	Smiths	0.25	0.45	0.25	0.05	25.22	27.11	41.39	71.69
11		Own brand	0.32	0.47	0.16	0.05	13.10	13.97	22.53	31.15
12		Red Deli	0.84	0.16			14.31	29.12		
13		Doritos	0.50	0.04	0.45	0.01	14.99	34.52	12.38	76.03
14	Confectionery	Cadbury	0.67	0.13	0.14	0.06	19.17	30.66	37.22	79.02
15		Own brand	0.21	0.73	0.06		20.61	15.31	50.54	
16		Mars	0.14	0.27	0.37	0.22	11.52	14.73	12.83	21.58
17		Lindt	0.75	0.25			12.38	27.99		
18		Allens	0.90	0.10			13.78	38.37		
19	Chilled desserts	Own brand	0.20	0.48	0.30	0.02	24.57	26.47	39.76	77.43
20		Yoplait	0.33	0.21	0.29	0.17	14.55	21.47	14.32	36.35
21		Vaalia	0.25	0.22	0.41	0.12	13.81	26.90	16.53	50.56
22		Chobani	0.20	0.64	0.16		14.04	17.23	35.58	
23		Jalna	0.32	0.55	0.13		13.94	16.43	32.24	
24	Cereal	Kelloggs	0.09	0.24	0.51	0.16	18.64	20.31	19.39	34.04
25		Uncle Toby	0.36	0.36	0.18	0.10	13.80	14.14	20.72	35.04
26		Sanitarium	0.60	0.31	0.09		18.43	20.51	38.01	
27		Own brand	0.70	0.30			13.48	28.12		

(3) Compute the brand price $price_b$ as a weighted average of prices of all n products from the brand b:

$$price_b = \sum_{m=1}^{n} \beta_m \, price_m$$

where $price_m$ is the price of product m, weighted by $\beta_m = N_m / N_b$. The term N_m is the number of days when customers bought m and $N_b = \sum_{m=1}^{n} N_m$ is the sum of number of days when customers bought products from b. The popular products have higher weights and play more important roles in the average price.

The daily brand price reflects the promotion time and the discount rate of that brand.

In terms of the other parameters, the number of segments for each brand is configured empirically based on the data fitness; the degree of the polynomial component of Eq. 5.8 is set to 2, which is adequate to capture the long-term patterns within the one-year observation period for our study.

Before analysing the behaviour patterns of each segment, we evaluate the accuracy of NHPP in modelling the purchase behaviour, and compare it with the baseline method HomoPP. We did not compare NHPP with other segmentation methods for two reasons. The first reason is that both NHPP and HomoPP are mixture models, which means that each customer has soft memberships in multiple segments instead of belonging to only one segment. The mixture models are more flexible in segmenting customers, so we did not include the traditional clustering techniques as baselines. The second reason is that our main purpose is to understand the effect of using the polynomial and periodic components in the intensity function, so comparing with the HomoPP which has a constant intensity function is a direct and effective approach. After evaluating NHPP's accuracy, we investigate the customer segments in depth, in order to understand the distribution of behaviour patterns for different categories and brands.

5.3.2 *Fitness of Purchase Events*

Table 5.2 lists the proportions and the estimated numbers of purchase events of all customer segments, obtained using the soft-membership customer segmentation introduced in Sect. 5.2. For each brand, the first three columns contain the unique brand ID, the category it belongs to and the brand name.

The name "own brand" refers to the supermarket's self-owned production brand. As we conduct a soft-membership segmentation, customers have a mixed membership over multiple segments for each brand, and the columns under "Size of Segment" are the proportions of all segments. The number of segments for each brand varies from 2 to 4; for the brand with fewer than 4 segments, the cells are empty for the non-existing segments. The last four columns display the estimated number of purchase events for that segment, which can provide an overview of the purchase rate of the segment. For example, the customers from segment 3 of Brand 1 purchased 29.31 times on average in 2014.

The plots in Fig. 5.4 show the cumulative number of purchase events for three sample customer segments. In each subplot, the blue line is the cumulative number of daily purchase events up to day t, which reflects the increments of purchase events. The red and black lines are the estimates produced by NHPP and HomoPP, respectively.

The blue lines demonstrate that the purchase rates were not stable over time for all these segments. In the top and middle subplots, the purchase rates were higher in the first half of the observation period, and decreased from the mid-year. In the bottom

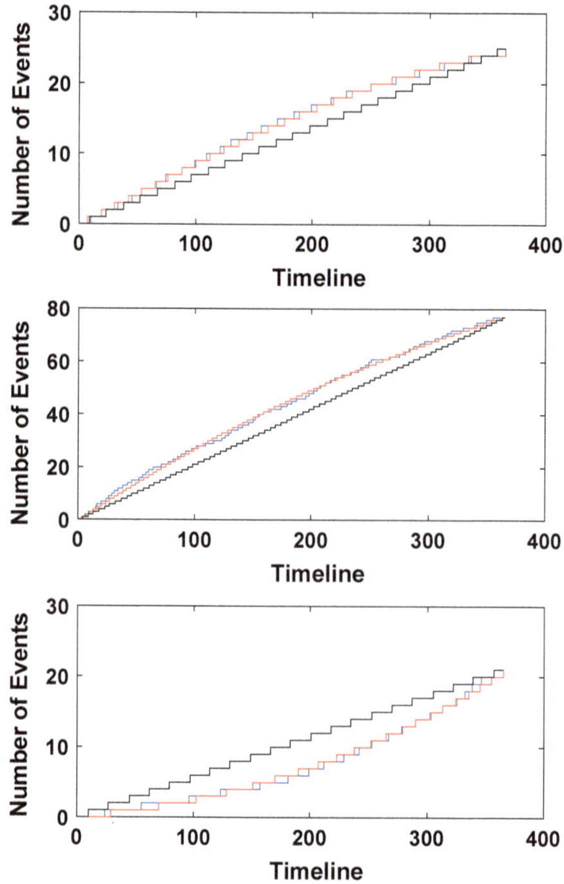

Fig. 5.4 Cumulative number of purchase events in three sample segments estimated by NHPP and HomoPP. Blue line: actual cumulative number of purchase events; red line: estimation by NHPP; black line: estimation by HomoPP. The time unit is day

subplot, the purchase rate increased through the whole year. Thus, the NHPP model is capable of accurately capturing the varying purchase rates for both scenarios.

To quantitatively evaluate our model, we first compute the absolute difference between the estimated and actual number of purchase events up to day t for each segment, which is illustrated as the gap between the blue line and the other two lines in Fig. 5.4. Then we average the differences over T days to compute the Mean Absolute Error (MAE):

$$\text{MAE} = \frac{\sum_{t=0}^{T} | \int_0^t \lambda(x)dx - N(t)|}{T} \tag{5.17}$$

The daily difference is computed, because we are not only interested in whether the model can estimate the total number of purchase events up to day T, but also want the model to capture the time of each purchase event. Based on the segment-wise

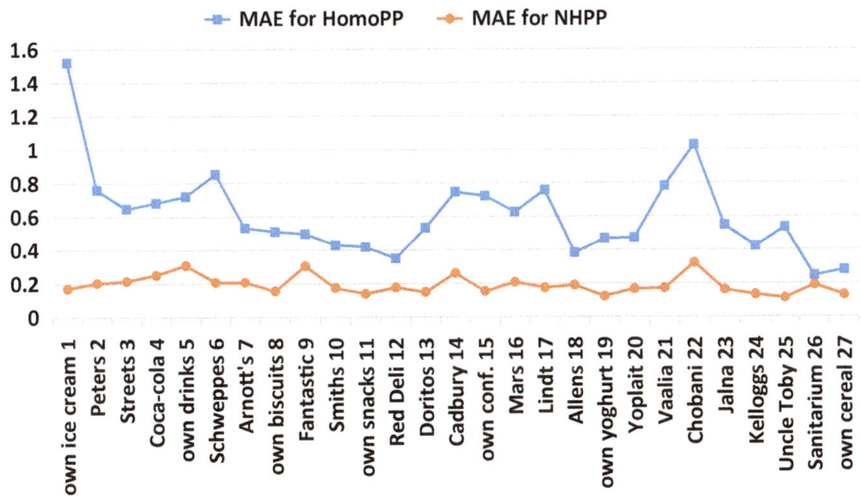

Fig. 5.5 Weighted MAE of HomoPP and NHPP models on all brands

MAE, we compute the weighted MAE across all segments of a brand, taking the segment sizes in Table 5.2 into consideration.

The weighted MAE of NHPP and HomoPP models for the 27 brands are shown in Fig. 5.5. Specifically, the MAE values in Fig. 5.5 are provided in Table 5.3. The results show that NHPP has lower MAE than HomoPP for all brands. The average MAE across all brands for HomoPP is 0.61, while it is 0.19 for NHPP. The paired t-test results validate that the NHPP model is *significantly* more accurate than the baseline model HomoPP ($p < 0.001$).

For some brands, e.g. Brands 1, 6 and 22, the MAE of HomoPP is particularly high, indicating that HomoPP cannot reliably model the dynamic purchase behaviour of these brands. From the category level, we find that the improvement of using NHPP (i.e. decrease of weighted MAE) ranges from 0.23 (Cereal) to 0.78 (Ice Cream), which indicates that the purchase rates of Ice Cream have larger fluctuations than of the other categories.

5.3.3 Long-Term and Short-Term Behaviour Patterns

From the visualisation of the purchase intensity function $\lambda_k(x)$, we notice that customer segments with similar number of purchase events can have significantly different long-term and short-term patterns.

For example, the purchase intensity curves of four segments (segments 1 and 3 of Brand 6, and segments 1 and 2 of Brand 8) are plotted in Fig. 5.6. Segments 1 and 3 of Brand 6 (in the upper row) have similar numbers of purchase events, which are 13.6 and 14.8. However, the long-term patterns for these two segments are different.

Table 5.3 Weighted segment-wise MAE values of HomoPP and NHPP for all products

ID	Category	Brand name	HomoPP	NHPP
1	Ice cream	Own brand	1.523	0.172
2		Peters	0.762	0.206
3		Streets	0.648	0.216
4	Soft drinks	Coca-Cola	0.684	0.252
5		Own brand	0.722	0.309
6		Schweppes	0.854	0.210
7	Biscuits	Arnott's	0.534	0.209
8		Own brand	0.511	0.156
9		Fantastic	0.497	0.306
10	Snacks	Smiths	0.431	0.175
11		Own brand	0.420	0.140
12		Red Deli	0.350	0.177
13		Doritos	0.533	0.150
14	Confectionery	Cadbury	0.748	0.263
15		Own brand	0.724	0.154
16		Mars	0.628	0.208
17		Lindt	0.759	0.174
18		Allens	0.383	0.189
19	Chilled desserts	Own brand	0.469	0.122
20		Yoplait	0.472	0.167
21		Vaalia	0.781	0.170
22		Chobani	1.026	0.321
23		Jalna	0.547	0.161
24	Cereal	Kelloggs	0.421	0.132
25		Uncle Toby	0.534	0.111
26		Sanitarium	0.244	0.189
27		Own brand	0.280	0.130

The purchase rate of segment 1 has a U-shape, which decreases in the first third of the year and then increases, while the purchase rate of segment 3 decreases steadily. Similarly, the subplots in the bottom row show that segments 1 and 2 of Brand 8 have different long-term patterns, but their numbers of purchase events are 19.2 and 19.3. Moreover, these two segments of Brand 8 also have slightly different periodic patterns, where the cycle length of segment 2 is shorter than the cycle length of segment 1.

Driven by the coefficients of the quadratic component $\sum_{d=0}^{D} w_d x^d$ $(d = 2)$ in Eq. 5.8, there are five major long-term patterns—*increase*, *decrease*, *U-shape*, *inverse U-shape* and *stable*. To determine the type of a pattern, we check two conditions:

(1) the direction of the parabola: upward ($w_2 > 0$) or downward ($w_2 < 0$);
(2) the location of the turning point $x_{tp} = -w_1/2w_2$.

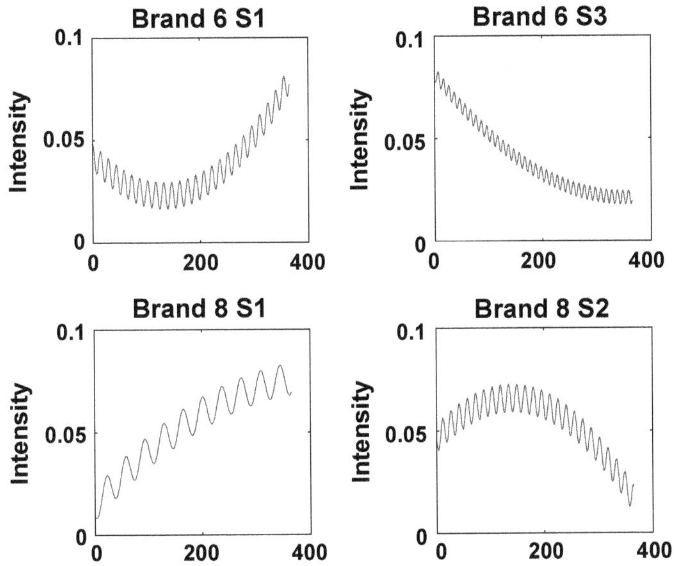

Fig. 5.6 Two pairs of segments of Brand 6 (top) and Brand 8 (bottom) with similar total number of purchase events but different temporal patterns. The horizontal axes are timelines (time unit: day) and the vertical axes are purchase intensity curves

Basically, if a parabola opens upward and the turning point is smaller than zero, the curve will be monotonic increasing for $x \geq 0$. In our analysis, we use a buffer of $\Delta t = 60$ days before the start and the end of the observation rather than the hard thresholds at 0 and T. If the turning point of an upward parabola is smaller than Δt, the curve will be defined as increase, since the decreasing part before the turning point is relatively small compared to the remaining time period. The specific conditions used to define five long-term patterns are as follows:

(1) Increase: $w_2 < 0$ & $x_{tp} > T - \Delta t$; $w_2 > 0$ & $x_{tp} < \Delta t$.
(2) Decrease: $w_2 < 0$ & $x_{tp} < \Delta t$; $w_2 > 0$ & $x_{tp} > T - \Delta t$.
(3) U-shape: $w_2 > 0$ & $\Delta t \leq x_{tp} \leq T - \Delta t$.
(4) Inverse U-shape: $w_2 < 0$ & $\Delta t \leq x_{tp} \leq T - \Delta t$.
(5) Stable: the difference between the minimum and maximum of the polynomial component is smaller than 0.01.

Based on these rules, we label the type of long-term pattern for each customer segment. Figure 5.7 demonstrates the representative curves for five long-term patterns. Each subplot only shows the polynomial component of the curve fitted for a segment.

Analyse Long-Term Patterns

In Table 5.4, we summarise the proportion of each long-term pattern to explore the trends shared at the category level. To compute this proportion, we first get the distribution of five long-term patterns for each brand. For instance, if a brand has

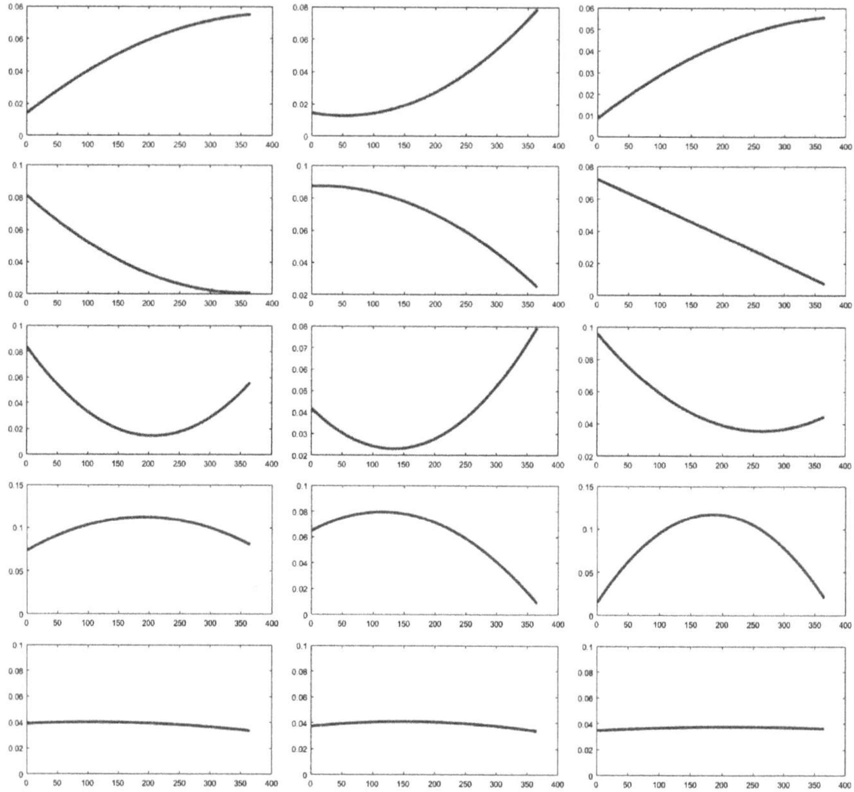

Fig. 5.7 Different types of long-term patterns (from top to bottom rows): increase, decrease, U-shape, inverse U-shape and stable. The horizontal axes are timelines ([0, 400], time unit: day). The vertical axes are purchase intensities ([0, 0.1])

Table 5.4 Distributions of five long-term patterns for different categories (the most frequent pattern for each category is in bold)

Category	Long-term patterns				
	Increase (%)	Decrease (%)	U-shape (%)	Inverse U (%)	Stable (%)
Ice cream	18.67	0	**42.67**	14.67	24.00
Soft drinks	11.67	9.00	**65.67**	13.67	0
Biscuits	10.00	0.67	0	**89.33**	0
Snacks	24.50	0	21.50	**33.00**	21.00
Confectionery	7.00	0	0	**93.00**	0
Chilled desserts	17.80	29.60	0	**52.60**	0
Cereal	0	21.25	2.25	**59.00**	17.50

two segments with weights of 0.7 and 0.3, the first segment is increase and the second is U-shape, the distribution of the five types (increase, decrease, U-shape, inverse U-shape and stable) is [0.7, 0, 0, 0.3, 0]. For the category level, we average the distributions over all relevant brands.

We first notice that two categories—Ice Cream and Soft Drinks, have large proportions of U-shape patterns, 42.67% and 65.67%, respectively. We also observe that the third category with U-shape patterns is Snacks (21.50%), which has a large gap from the top two. The other categories have very few segments with U-shape patterns. Considering the context, this result makes sense as it is winter in the mid-year in Australia. It is reasonable that customers buy less ice cream and soft drinks during winter than summer, which leads to U-shape purchase rate curves.

As for inverse U-shape patterns, the Confectionery Category ranks first, with 93% inverse U-shape patterns and only 7% increase patterns. The Biscuits Category has a similar distribution, with 89.33% inverse U-shape patterns and only 10.67% increase and decrease patterns. These two categories are followed by Cereal (59%) and Chilled Desserts (52.6%), although there is a substantial gap between these categories and the top two. The results indicate that a large portion of customers tend to purchase more energy-rich foods like confectionery and biscuits in winter rather than in summer.

For the other three types of long-term patterns, the proportions are lower than 30%, without clearly dominant patterns. The Snacks Category has a roughly even distribution across the four patterns which reflects the distinct behaviour of different segments, influenced by seasons and personal preference drifts. The Ice Cream, Snacks and Cereal have about 20% of stable patterns. This means that these customer segments have relatively stationary purchase rates and the customers buying them are less sensitive to seasonal effects.

Analyse Short-Term Patterns

As for the short-term periodic patterns, we analyse the cycle lengths of the purchase rate curves. We did not analyse the amplitude of the sine component, as the difference between amplitudes were relatively small compared to the magnitude of the polynomial component. The minimum, maximum and mean of cycle lengths for each category are illustrated in Fig. 5.8. Similarly to previous measures, the mean value for each category is the average of segment-wise cycle lengths, weighted by the segment sizes. The weighted mean cycle length for Cereal is 25.4, almost double the mean cycle lengths of the other categories. The main reason for this is that cereals are often sold in large packs (e.g. 1 kg), which take longer to consume. For the other categories, the mean cycle lengths are between 12.9 and 15.9 days, which are all about 2 weeks. Since the segments with maximum or minimum cycles generally have smaller segment sizes, the weighted means are all about 2 weeks.

The analysis of long-term and short-term patterns demonstrates the differences between the customer behaviour for different categories. Categories which are influenced by seasonal effects more strongly, have more U-shape and inverse U-shape patterns; whereas other categories have roughly even distributions of five long-term patterns. Additionally, the categories which are easy to store and consumed slower, such as Cereal, have longer purchase cycles.

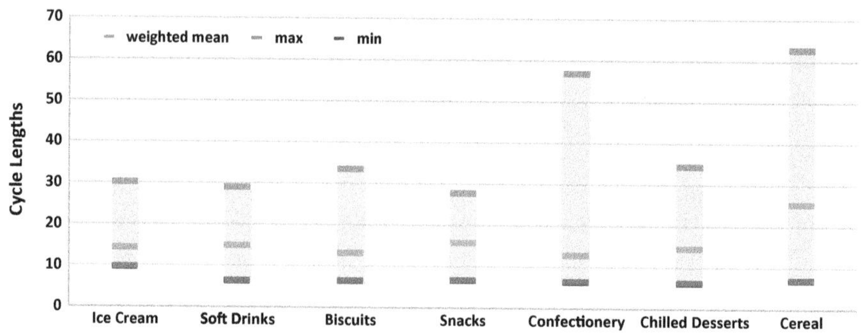

Fig. 5.8 Cycle lengths (time unit: day) of short-term periodic purchases for different categories

5.3.4 Customer Responses to Promotions

To analyse how a customer segment responds to the promotions of a brand, we compute the correlation between the segment-level purchase rates and the daily brand price. The correlation values range in $[-1, 1]$, and negative correlations are expected if the customer behaviour of that segment is receptive to the price reduction.

Table 5.5 reports the correlation analysis results for all brands and customer segments. Strong negative correlations (≤ -0.2) are highlighted in bold. The second last column lists the proportion of customers who are receptive to price reductions, which is computed by summing the sizes of all the segments with negative correlations. The Mean Discount Rate (MDR) in the last column is computed by averaging $(1 - daily\_price)$, because the daily price is relative to the maximal price, so $(1 - daily\_price)$ reflects the daily discount rate.

The results show that Brands 2, 3, 6, 7, 16, 17, 18 and 26, have negative correlations for all their customer segments, so that the proportions of customers who are receptive to promotions are 100% (highlighted in bold) for these brands. These brands also have higher average MDR, which is 0.25, while the average MDR for all brands is 0.19. This means that these brands have larger or more frequent sales promotions than the other brands. Therefore, having more customers receptive to the price reductions in these brands is beneficial for the supermarket.

There are 7 supermarket's own brands (Brands 1, 5, 8, 11, 15, 19 and 27), one for each category. The MDR of all these brands are the lowest in their corresponding category. The average MDR for these supermarket's brands is 0.1, which is about half of the overall average for all brands, 0.19. For the supermarket's own brands, the average proportion of customers who are receptive to promotions is 43.8%, while for the other brands it is 70.1%. The substantial difference observed between these two proportions confirms that fewer customers buying supermarket's own brands are receptive to price reductions, compared to the customers buying other brands. One of the possible reasons is that the supermarket's own brands have less attractive discounts, which has been illustrated by the low MDR, and their prices are generally

Table 5.5 Customer responses to price reduction and mean discount rates for all brands. Strong negative correlations (≤ -0.2) and the brands with all customers receptive to promotions are highlighted in bold

ID	Category	Brand name	Correlation with price				Receptive to promotions	Mean discount rate
			1	2	3	4		
1	Ice cream	Own brand	**−0.20**	0.35	−0.14		0.72	0.12
2		Peters	−0.06	**−0.20**	−0.11		**1.00**	0.25
3		Streets	−0.01	**−0.20**			**1.00**	0.24
4	Soft drinks	Coca-Cola	0.41	−0.12	**−0.28**	0.08	0.64	0.23
5		Own brand	0.26	**−0.62**	0.54	0.26	0.25	0.16
6		Schweppes	**−0.31**	**−0.23**	−0.17	−0.06	**1.00**	0.27
7	Biscuits	Arnott's	−0.03	−0.09	−0.08	−0.10	**1.00**	0.22
8		Own brand	−0.19	0.25	0.02	0.18	0.31	0.09
9		Fantastic	0.15	−0.08	0.26		0.42	0.28
10	Snacks	Smiths	0.32	**−0.38**	0.26	0.08	0.45	0.23
11		Own brand	0.16	**−0.37**	0.47	−0.04	0.52	0.09
12		Red Deli	0.11	0.05			0.00	0.25
13		Doritos	−0.14	−0.03	0.03	0.00	0.55	0.25
14	Confectionery	Cadbury	−0.04	0.04	0.06	0.02	0.67	0.29
15		Own brand	0.09	−0.12	−0.04		0.79	0.08
16		Mars	−0.03	−0.08	−0.06	−0.04	**1.00**	0.29
17		Lindt	−0.12	**−0.20**			**1.00**	0.27
18		Allens	−0.18	−0.16			**1.00**	0.22
19	Chilled desserts	Own brand	0.16	−0.003	0.10	0.08	0.48	0.11
20		Yoplait	−0.15	0.19	0.15	−0.03	0.50	0.17
21		Vaalia	0.18	0.05	−0.17	−0.01	0.54	0.15
22		Chobani	−0.11	0.10	−0.07		0.36	0.19
23		Jalna	**−0.52**	0.39	0.18		0.32	0.12
24	Cereal	Kelloggs	**−0.20**	0.13	−0.19	−0.18	0.76	0.16
25		Uncle Toby	**−0.34**	−0.12	0.06	**−0.27**	0.82	0.14
26		Sanitarium	−0.07	−0.03	−0.03		**1.00**	0.18
27		Own brand	0.02	0.26			0.00	0.07

lower than the other brands, so that the purchase rates do not have strong negative correlations with the price.

From the category level, Ice Cream and Confectionery have larger proportions of customers who are receptive to price change, which are 91% and 89%, respectively. Snacks and Chilled Desserts only have 38% and 44% of customers who are receptive to promotions, respectively, which is less than half of the proportions for Ice Cream and Confectionery. There are two possible reasons: firstly, the average MDRs of Ice

Cream and Confectionery are 0.2 and 0.23, which are higher than Chilled Desserts, 0.148; secondly, products of Snacks (e.g. potato chips and nuts in small packs) are generally cheaper than Ice Cream and Confectionery (e.g. chocolates).

5.4 Summary

In this chapter, we propose to use a mixture of NHPP models to segment customers in terms of latent long-term and short-term purchase behaviour patterns. We conduct a case study on 27 brands from food categories and contribute from three aspects:

- We validate the significant advantages of using our NHPP model to estimate the occurrence of purchase events, as the purchase behaviour is non-stationary over time.
- We decompose the dynamic behaviour patterns into long-term trends related to factors such as seasonal effects and preference drifts, and short-term periodic purchase driven by personal needs and promotions.
- We explore customer responses to promotions using the correlation between purchase rates and price, and identify customers who are receptive to price reductions.

Our comprehensive analysis of segmentation results identifies typical behaviour patterns shared by customers and investigates the behaviour pattern distributions at both brand and category levels.

We find that most long-term patterns seem to be caused by seasonal effects and personal preference changes. If the demand of a category tends to be sensitive to the season, the category will have more U-shape and inverse U-shape patterns. If the purchase behaviour of a category is mainly driven by customer preferences, different types of long-term patterns are more evenly distributed, which reflects the preference drifts of different customers. The short-term patterns are determined by regular promotions, shopping habits and the necessity of products. Another influential factor for the short-term pattern is whether the product is easy to stockpile at home, which has been validated by the longer cycle length of cereals.

As for the receptiveness of a customer segment to price changes, we analyse the correlation between purchase rates and product price. If there are regular attractive promotions, the purchase rate is negatively correlated with the price. In our case study, the supermarket's own brands do not have frequent promotions but have lower prices than their branded counterparts, so there are fewer negative correlations for these brands. We find that higher proportions of customers have been motivated by the larger price reductions, and there are 8 brands with all segments receptive to promotions. Based on the amount of customers whose purchase rates are affected by price, the business management could examine the impact of their promotions and target customers with better marketing strategies.

From the practical viewpoint, the analysis of behaviour patterns and customer responses to promotions could help the business management to select suitable rates of promotions. In addition, our method could compare customer behaviour for different categories, according to the distributions of long-term patterns and the cycle lengths of short-term patterns. This provides a guide to optimising the promotion strategies for customers of different categories. Furthermore, the customer segmentation based on behaviour patterns supports target marketing, as it helps stakeholders to get a clear understanding of their market and target appropriate customers.

The current method can detect behaviour patterns and evaluate the program impact. In future work, we could consider to extend this temporal modelling framework by adding the prediction of customer behaviour patterns into it. The prediction of behaviour patterns can further improve the accuracy of recommendation and effectiveness of resource management.

References

1. Ehrenberg ASC (1959) The pattern of consumer purchases. Appl Stat 26–41
2. Chatfield C, Goodhardt GJ (1973) A consumer purchasing model with Erlang inter-purchase times. J Am Stat Assoc 68(344):828–835
3. Morrison DG, Schmittlein DC (1988) Generalizing the NBD model for customer purchases: what are the implications and is it worth the effort? J Bus Econ Stat 6(2):145–159
4. Trinh G, Rungie C, Wright M, Driesener C, Dawes J (2014) Predicting future purchases with the poisson log-normal model. Mark Lett 25(2):219–234
5. Bucklin RE, Gupta S (1992) Brand choice, purchase incidence, and segmentation: an integrated modeling approach. J Mark Res
6. Bucklin RE, Gupta S, Siddarth S (1998) Determining segmentation in sales response across consumer purchase behaviors. J Mark Res 189–197
7. Chen M-C, Chiu A-L, Chang H-H (2005) Mining changes in customer behavior in retail marketing. Expert Syst Appl 28(4):773–781
8. Huang C-K, Chang T-Y, Narayanan BG (2015) Mining the change of customer behavior in dynamic markets. Inf Technol Manag 16(2):117–138
9. Allenby GM, Leone RP, Jen L (1999) A dynamic model of purchase timing with application to direct marketing. J Am Stat Assoc 94(446):365–374
10. Adamopoulos P, Todri V (2015) The effectiveness of marketing strategies in social media: evidence from promotional events. In: Proceedings of the 21th ACM SIGKDD international conference on knowledge discovery and data mining. ACM, New York, pp 1641–1650
11. Iwata T, Watanabe S, Yamada T, Ueda N (2009) Topic tracking model for analyzing consumer purchase behavior. In: Proceedings of the 22nd international joint conference on artificial intelligence. AAAI Press, California, pp 1427–1432
12. Kim H, Takaya N, Sawada H (2014) Tracking temporal dynamics of purchase decisions via hierarchical time-rescaling model. In: Proceedings of the 23rd ACM international conference on information and knowledge management. ACM, New York, pp 1389–1398
13. Sheldon M (1996) Ross, stochastic processes, vol 2. Wiley, New York
14. Ferraz Costa A, Yamaguchi Y, Juci Machado Traina A, Traina Jr C, Faloutsos C (2015) RSC: mining and modeling temporal activity in social media. In: Proceedings of the 21th ACM SIGKDD international conference on knowledge discovery and data mining. ACM, New York, pp 269–278
15. Bishop CM (2006) Pattern recognition and machine learning. Springer, New York
16. Dempster AP, Laird NM, Rubin DB (1977) Maximum likelihood from incomplete data via the EM algorithm. J R Stat Soc Ser B (Methodol) 39(1):1–38
17. Ng A (2016) The EM algorithm. In CS229 Lecture notes

Chapter 6
Evaluating Impact of the Web-Based Health Program

6.1 Background

Numerous websites and mobile applications have been created for health and fitness, which can support users to easily manage diet plans, count calorie intake, keep activity journals and communicate with others. The behaviour change programs increasingly utilise websites, mobile applications and wearable devices to maximise the program benefits and motivate participants to achieve their goals. For example, an Australian supermarket recently run a program that encouraged people to do more physical activity.[1] By linking activity trackers with loyalty cards, participants can get discounts on fresh products once reaching the milestones, e.g. 100,000 steps. These technologies can provide interactive and personalised services to participants and allow program stakeholders to manage and analyse behavioural data.

After collecting behavioural data, the program stakeholders often want to evaluate if the program is effective, and whether the supporting websites and applications are helpful, so that they can improve the program to better motivate participants and benefit a broader population. Previous studies mainly used descriptive statistics, rating-based surveys and statistical tests to evaluate the impact of programs and supporting web applications [1–3]. Although these methods can evaluate the overall impact of the health programs, they can not track the behaviour changes of individual participants over time and compare the program impact on the participants with different types and levels of engagement with the web applications.

Problem Formulation and Organisation

As described in Chap. 2, our supermarket health program, HealthierU, was designed to motivate participants to adopt a healthier lifestyle and improve their eating habits. There are two components in our health program: a supporting website with interactive tools for the diet program and 10% discount when purchasing fresh products in the supermarket. Our research tasks are to:

---

[1] https://www.flybuys.com.au/collect#/health/stepstosavings.

L. Luo, *Temporal Modelling of Customer Behaviour*,
Springer Theses, https://doi.org/10.1007/978-3-030-18289-2_6

(1) evaluate if the supporting website is helpful for the health program;
(2) measure the impact of the health program on different types of participants.

We construct reliable individual purchase rate curves, which act as the backbone of the program impact analysis. To overcome overfitting problems, the individual purchase rate curve is modelled as a mixture of purchase behaviour models for latent customer segments, weighted by the individual soft membership in all segments. We design four criteria for partitioning participants by their levels of interactions with the program website. Then we quantify the impact of the health program on participants based on their purchase behaviour changes after joining the program.

This chapter is organised as follows. This section introduces the main contributions, related work of this chapter and its relationship with Chap. 5. Section 6.2 describes three modules of our methodology to evaluate the impact of the health program. Section 6.3 explains the details of our case study on the supermarket health program and analyses the participants' responses to the program based on their interactions with the program website and their purchase rates of different products. Section 6.4 summarises our findings and contributions of this chapter.

Contributions

Our case study on the supermarket health program compares the program impact on participants with different levels of interactions with the program website. Through the case study, we establish that:

(1) During the health program, the participants increased their purchase of fresh products.
(2) The distributions of behaviour patterns are different for participants with different levels of interactions with the program website.
(3) The impact of the health program is stronger for the participants with higher levels of interactions with the program website.

Related Work

Existing studies have analysed the effects of using support websites and applications in the health program. Wieland et al. [4] showed that the digital systems provided an effective support for weight loss and maintenance compared to the offline interventions. Brindal et al. [2] demonstrated the positive role played by the support applications in their meal replacement program. However, only using descriptive statistics of the weight loss and surveys can be subjective and insufficient to capture how participants' behaviour gradually changes due to the program participation. Tracking behaviour changes can provide valuable information to evaluate the impact of a program objectively [5].

Given the context of the supermarket health program, we build purchase behaviour models and measure the changes in purchase rates of different products. Considering the accuracy and interpretability, we use stochastic processes, a mixture of NHPP models introduced in Chap. 5 as the base of our purchase behaviour model. It is worth noting that the model in Chap. 5 focused on the customer segmentation and segment-level responses to promotions, whereas in this chapter we have a different purpose, new methodology and evaluation. We construct individual purchase rate curves based on the segment-level model and propose criteria for partitioning customers, which allow us to track individual behaviour changes and evaluate the impact of our program and its supporting website.

6.2 Methodology

The health program stakeholders desire to identify which participants are motivated by the program to buy more fresh products, and whether the website can effectively trigger stronger behaviour changes towards a healthier lifestyle. To this end, we take both the online and offline components of the health program into consideration and explore the program impact on different types of customers, who are grouped based on their interaction with HealthierU, such as website logon, number of diaries they wrote and the usage of the weight tracker (the groups will be detailed in Sect. 6.2.2).

Our method consists of three modules as shown in Fig. 6.1, where we:

(1) *Model purchase behaviour* using a mixture of NHPP, based on transaction records;
(2) *Model the individual purchase rates* of different products for all customers;
(3) *Evaluate the program impact* by considering both the usage of HealthierU website and the fluctuations of individual purchase rates during the health program.

This section describes the construction of individual purchase rate curves and our solutions to three questions of evaluating the program impact:

Q1: How do the purchase rates during the program differ from those observed before and after the program?
Q2: How does the purchase rate change during the program?
Q3: Are the active users of the HealthierU website more receptive to the program than the others?

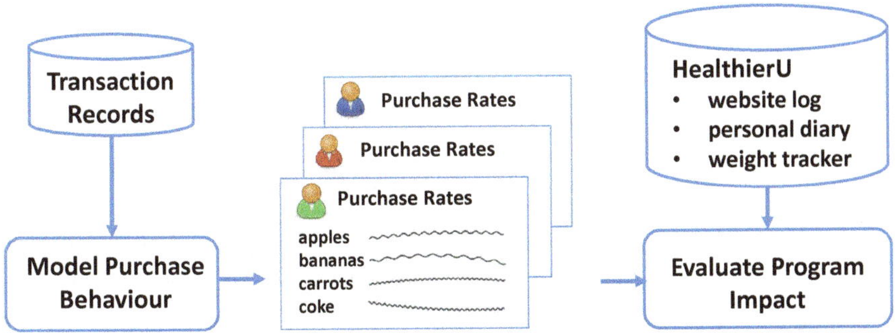

Fig. 6.1 Flowchart of our method

6.2.1 Model Individual Purchase Rates

Our first task is to accurately model individual purchase behaviour. To prevent overfitting the sparse individual purchase records, we include additional information to regularise the individual models. Thus, we segment customers by constructing a mixture of NHPP models to identify typical behaviour patterns from the purchase events of all customers. After that, we consider each customer as a member of multiple segments with soft membership, and the individual purchase rate curve can be computed by the weighted sum of segment-level models.

The advantages of modelling individual purchase rate via a mixture of NHPP are:

(1) It can mitigate the influence of sparsity and noise in the actual purchase records.
(2) The individual purchase rate curve can be easily interpreted as a weighted combination of typical latent purchase behaviour patterns shared by all customers.
(3) It can show the individual purchase behaviour changes and facilitate the analysis of program impact.

Given a transaction dataset with U customers and M products, for a product m, we aim to construct individual purchase rate curve for all customers buying m, $\{u_i\}^{(m)}$ $(i \in \{1, 2, ..., U\})$. We first adopt the mixture of NHPP defined in Chap. 5 with the following intensity function (same as Eq. 5.8):

$$\lambda(x) = \sum_{d=0}^{D} w_d x^d + a \sin(bx + c) \tag{6.1}$$

The EM algorithm is used to infer the coefficients of segment-level NHPP model $\mathbf{\Theta}$, segment sizes $\mathbf{\Phi}$ and soft membership for all customers $\mathbf{\Pi}$.

Construct Individual Purchase Rate Curve

The individual purchase rate curve $\lambda_i(x)^{(m)}$ represents the estimated daily purchase rate of u_i buying product m. We omit m in the following descriptions, as they are all in the context of a certain product m. Our individual purchase rate curve is constructed based on the customer segmentation result as follows:

$$\lambda_i(x) = \sum_{k=1}^{K} \pi_{ik} \lambda_k(x) \tag{6.2}$$

which is a linear combination of the purchase behaviour models for K segments, weighted by u_i's soft membership in each segment. The polynomial component of $\lambda_i(x)$ can capture behaviour patterns such as increase, decrease, U-shape and inverse U-shape, depending on the sum of polynomial coefficients of all segments, $w_d = \sum_{k=1}^{K} \pi_{ik} w_{dk}$.

Consider the example in Fig. 6.2, where there are three customer segments with increase, decrease and inverse U-shape patterns, respectively. Assuming a customer

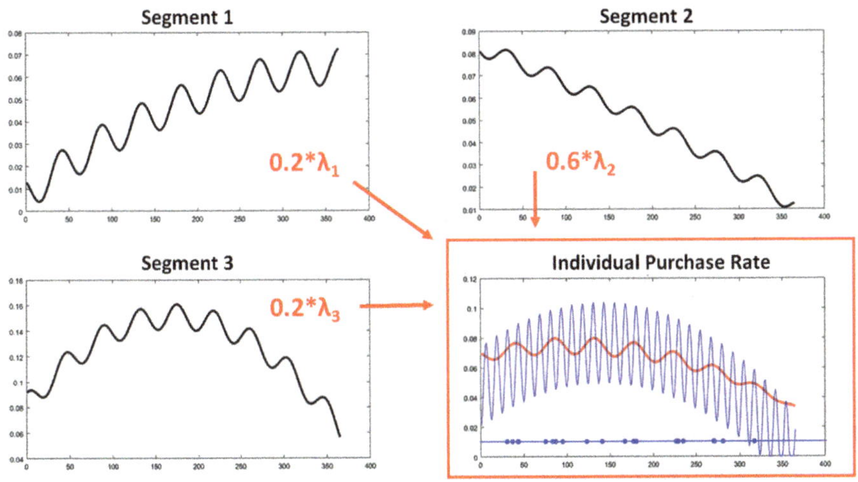

Fig. 6.2 Using sparse individual records (blue circles) can cause overfitting problems (as shown by the blue curve), while our purchase rate curve (red curve) formed by the weighted sum of segment-level models (black curves) is more reliable

u_i has soft membership $\pi_i = [0.2, 0.6, 0.2]$ in these segments, the individual purchase rate would be the red curve in the bottom right plot. Compared to the blue curve which fits u_i's sparse purchase records (blue circles) directly, our model can overcome overfitting problems and reveal the underlying purchase behaviour patterns.

Given $\lambda_i(x)$, we can obtain the customer purchase rate on any day, which allows to analyse the changes of their behaviour. Most importantly, the derivative of $\lambda_i(x)$ can reflect the rate of change of u_i's purchase behaviour. Moreover, the integral of $\lambda_i(x)$ over a time period shows the estimated number of purchases u_i would make over that period.

6.2.2 *Evaluate Program Impact*

As mentioned above, the purchase rate curve $\lambda_i(x)$ can facilitate the evaluation of the program impact, so we utilise $\lambda_i(x)$ to answer the following three questions.

Q1: How do the purchase rates during the health program differ from those observed before and after the health program?

We firstly obtain the mean purchase rates in three phases of the observation period: before, during and after the health program. The mean purchase rate r_{ph} of phase ph is computed by:

$$r_{ph} = \frac{1}{t_2 - t_1 + 1} \int_{t_1}^{t_2} \lambda_i(x)dx \tag{6.3}$$

where t_1 and t_2 are the start and end time of ph. Then, we conduct statistical tests to check if the purchase rates across three phases are significantly different. The test we use is the one-way ANOVA [6], which is an extension of the t-test to compare more than two groups based on one independent variable, the mean purchase rate in our case. Our hypothesis is that the purchase rates of fresh products during the health program would be higher than the other two phases.

Q2: How does the purchase rate change during the health program?

To gain insights of the behaviour changes during the health program, we analyse the behaviour patterns of different customer segments and the distribution of various patterns for all customers. Our hypothesis is that the active participants of the health program would have more significant purchase rate changes during the program, due to the program impact.

The coefficients of $\sum_{d=0}^{D} w_d x^d$ in Eq. 6.1 can capture different patterns of behaviour changes. Setting d=2, based on whether the parabola opens upward ($w_2 >$ 0) or downward ($w_2 < 0$) and the location of the turning point $x_{tp} = -w_1/2w_2$, we observe five long-term patterns: *increase*, *decrease*, *U-shape*, *inverse U-shape* and *stable*. More details about the specific conditions used to determine different types of long-term patterns are given in Sect. 5.3.3. For any product, the distribution of these long-term patterns can show the trend of customer preferences and indicate the proportion of customers who have been influenced by the health program.

Q3: Are the active users of the HealthierU website more receptive to the health program than the others?

this chapter, being receptive to the health program refers to increasing the purchase rate of fresh products. The main challenge of this question is that the purchase behaviour could also be impacted by other factors such as promotions and seasons, due to the variations of price and product availability among different seasons. Particularly, the purchase of fresh products may be significantly affected by changes of seasons. For example, the purchase rate of grapes may decrease in winter for most customers during the program. We desire to identify which customers have larger increases in their purchase rates compared to others. Our hypothesis is that the customers who are the active users of the HealthierU website are more receptive to the health program than the others. Therefore, we evaluate the program impact in two steps.

In the first step, we use the derivative of $\lambda(x)$ to measure the rate of changes in the purchase rate curve. The positive derivative means that the purchase rate was increasing at that time, and the negative derivative means decreasing. To evaluate the increase of the purchase rate during the health program for u_i, we define a measure σ_i as:

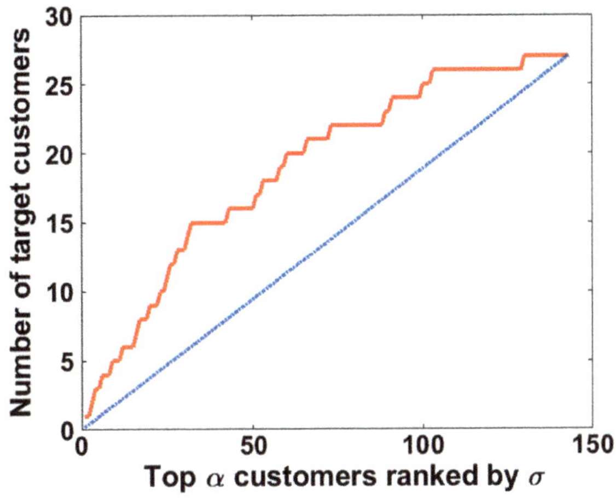

Fig. 6.3 Illustration of the impact score

$$\sigma_i = \sum_{x \in S} \lambda_i'(x), \;\; S = \{x|\lambda_i'(x) > 0, x \in \{t_1, t_1 + 1, ..., t_2\}\} \tag{6.4}$$

which is the sum of all positive $\lambda_i'(x)$ from t_1 to t_2 for u_i. The measure σ_i reflects both the duration and rate of the increases of the purchase rate during the health program.

In the second step, we rank all σ_i in a descending order, so that the high-ranked customers have a larger increase in their purchase rates. We examine if certain groups of customers, e.g. those who used the website frequently or those who lost more weight, are more receptive to the health program. Specifically, the group of customers we are interested in is labelled as the *target group* (i.e. treatment group), and we compare their ranks to other customers (i.e. control group). Therefore, we would like to check if the customers in the target group have higher ranks, which means larger increases in their purchase rates, than the others.

In Fig. 6.3, we show the number of target customers among top α customers sorted by σ for each $\alpha \in \{1, \ldots, U\}$. The blue line corresponds to the baseline, when the target customers distribute uniformly among all customers. The area under this blue line is 0.5. The red line shows the actual number of target group customers among top α customers. For example, the plateau of the line between $\alpha = 32$ and $\alpha = 42$ means that all the customers who ranked between 32 and 42 are not from the target group. We use the area under the red line to measure the program impact quantitatively. Formally, the *impact score* of the health program with respect to a target group is defined as:

$$impact = \frac{1}{U_{target} U} \sum_{\alpha=1}^{U} n_\alpha \tag{6.5}$$

where n_α is the number of customers in the target group and also in the set of top-α customers with the highest σ; $\sum_{\alpha=1}^{U} n_\alpha$ is the sum of n_α for each $\alpha \in \{1, \ldots, U\}$;

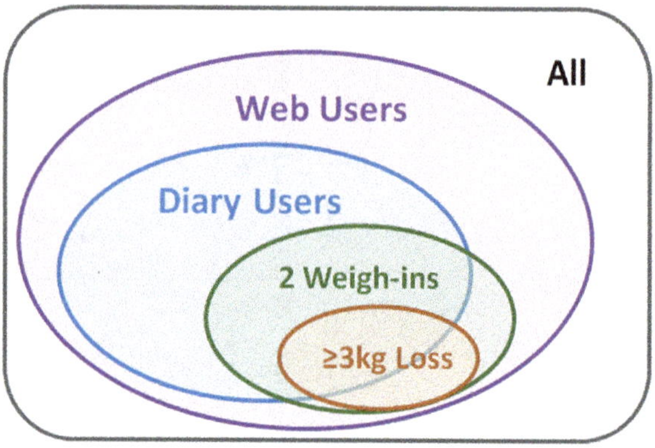

Fig. 6.4 Relationship of target groups formed by four partition methods

U_{target} is the number of customers in the target group; and U is the total number of customers. The principle of *impact* is similar to the measure – Precision at k ($P@k$) [7].

The baseline of *impact* is 0.5 and the greater *impact* values correspond to lines above the blue line in Fig. 6.3. The greater *impact* value implies that the ranks of the customers in the target group are higher than the others, indicating that the health program is more effective for the target group.

When analysing purchase pattern distributions and impact scores, we partition customers into exclusive groups in four ways, considering the frequency and type of their interactions with HealthierU website. Our goal is to compare whether the customers from different groups[2] have different purchase behaviour for fresh products and other categories. The four ways to partition the customers are:

PT1 *active versus inactive web users*: if the customer used HealthierU 12 times or more during the program;

PT2 *active versus inactive diary users*: if the customer wrote 12 diary entries or more during the program;

PT3 *2 versus* $<$ *2 valid weigh-ins*: if the customer reported weight at least twice and the reports were at least 6 weeks apart;

PT4 $\geq$*3 kg versus* $<$*3 kg weight loss*: if the customer with at least 2 valid weight reports lost more than 3 kg.

These four partitions are illustrative examples and other ways may be considered. These thresholds are set to generate comparable groups and prevent impact of the noise. The relationship among the target groups formed by four different partition methods are illustrated in Fig. 6.4. The specific sizes of customer groups, e.g. active

[2]In this chapter, partitioning based on the website usage and customer segmentation based on purchase behaviour (as in Sect. 6.2.1) are different concepts. We use the term 'customer segment' for the behaviour-model-based segmentation result, and the term 'customer group' for the rule-based partitioning result.

diary users, are different for different products. The partitions enable us to compare the program impact on the customers with different levels of interactions with HealthierU.

6.3 Case Study

We select 38 most popular products as listed in Table 6.1 based on the purchase records. There are 21 types of fresh fruits and vegetables and 17 product brands[3] from other categories such as soft drinks and confectionery. The own brand refers to the supermarket's self-owned production brand.

The health program started on May 26 and ran for 24 weeks. The participants were able to access the HealthierU website at any time during the health program. They were required to complete two surveys of basic information and health conditions. The first survey was before the start of the health program and the second one was 12 weeks after the start, at the end of the first cycle. The participants who did not complete the second survey were not eligible for the second cycle, so the program duration for them was 12 weeks. Among 931 participants, 198 customers attended both cycles of the program.

For all participants, the observation period is split into three phases as follows:

(1) *before* the program (from January 1 to May 25);
(2) *during* the program (from May 26 to the end of the program determined by the completion of the second survey);
(3) *after* the program (from the end of the program to December 31).

For each product, the active customers are those who bought the product more than 10 times. The construction of NHPP model is based on the active customers of each product. In terms of the other parameters, the number of segments for each product is configured empirically based on the data fitness; the degree of the polynomial component is 2, which is the same as what we have used in Sect. 5.3.1.

6.3.1 *Purchase Rate Analysis*

We apply the mixture of NHPP to the customer segmentation. Table 6.1 provides an overview of the segmentation results, which shows the estimated total number of purchases for each customer segment. For each product, the number of segments is 3 or 4, where the empty cell means that the product does not have that segment.

We conduct ANOVA tests to check if the mean purchase rates *before*, *during* and *after* the program are different. The *p-value* and the mean purchase rates in three phases of all products are listed in Table 6.2.

---

[3] They are also referred to as 'product' in our case study for simplicity.

Table 6.1 Product information and the estimated number of purchases for customer segments generated by NHPP models

ID	Category	Product name	Estimated *N* for segments			
			S1	S2	S3	S4
1	Vegetables	Carrots	14.35	14.64	34.84	24.23
2		Cup mushrooms	13.49	15.57	22.4	43.34
3		Red onions	15.87	14.45	12.12	32.06
4		Sweet potatoes	50.21	14.66	17.09	29.03
5		Red capsicum	13.4	20.9	28.1	46.34
6		Broccoli	12.86	17.98	29.8	41.48
7		Cucumbers	13.73	14.35	24.52	45.1
8		Zucchini	14.18	16.7	33.62	81.31
9		Field tomatoes	12.79	15.45	18.89	33.83
10		Lettuce	14.44	12.53	21.48	39.28
11		Truss tomatoes	13.91	14.15	29.48	57.41
12		Leb. cucumbers	13.22	15.83	19.07	46.6
13		Grape tomatoes	13.44	49.81	21.27	112.09
14		Spring onions	13.53	13.8	19.95	33.16
15		Mix salad	18.22	15.99	13.99	41.03
16	Fruits	Apples	31.64	15.48	16.77	58.11
17		Bananas	15.91	29.44	46.78	77.26
18		Strawberries	15.64	19.8	37.21	64.03
19		Avocado	15.52	14.68	26.21	41.47
20		Blueberries	28.24	12.85	14.45	43.07
21		Grapes	12.28	12.76	14.26	18.36
22	Soft drinks	Coca-Cola	15.28	37.05	73.39	159.69
23		Own drinks	26.01	15.65	44.84	94.22
24		Schweppes	15.16	14.77	41.03	
25	Biscuits	Arnott’s	29.2	15.18	46.37	84.24
26		Own biscuits	13.53	15.12	29.09	67.51
27	Snacks	Smiths	15.04	15.04	32	58.51
28		Own snacks	12.41	17.79	13.53	22.87
29		Doritos	48.5	12.39	14.63	
30	Confectionery	Cadbury	13.25	17.22	33.05	75.57
31		Own confectionery	15.61	50.77	18.78	
32		Mars	12.07	13.51	23.81	
33	Yoghurt	Own yoghurt	23.6	14.03	34.85	77.33
34		Yoplait	14.63	18.39	19.43	35.13
35		Vaalia	16.23	18.22	45.47	
36	Cereal	Kelloggs	14.83	13.39	17.34	32.08
37		Uncle Toby	20.8	12.94	14.47	35.84
38		Sanitarium	13.88	14.38	19.7	37.5

Table 6.2 ANOVA test results of the mean purchase rates before, during and after the program for all products. Statistical significant *p-values* ($p < 0.05$) and the corresponding product names are in bold. The significant *p-values* of fresh products with larger rates in Phase 2 are underlined

ID	Category	Product name	*p*	Mean purchase rates		
				Phase 1	Phase 2	Phase 3
1	Vegetables	Carrots	0.102	0.0202	0.0223	0.0206
2		Cup mushrooms	0.16	0.0167	0.0169	0.0152
3		Red onions	0.583	0.015	0.0141	0.0145
4		**Sweet potatoes**	**<u>0.022</u>**	0.0124	0.0139	0.012
5		Red capsicum	0.184	0.024	0.0224	0.0218
6		**Broccoli**	**<u>0.000</u>**	0.0201	0.0242	0.0199
7		Cucumbers	0.181	0.0189	0.018	0.02
8		**Zucchini**	**<u>0.007</u>**	0.016	0.0179	0.0146
9		**Field tomatoes**	**<u>0.000</u>**	0.014	0.0175	0.0155
10		Lettuce	0.186	0.0133	0.0119	0.0126
11		**Truss tomatoes**	**0.001**	0.015	0.012	0.0132
12		Leb. cucumbers	0.067	0.014	0.0124	0.0119
13		Grape tomatoes	0.062	0.0138	0.0121	0.0115
14		Spring onions	0.617	0.0112	0.0107	0.0108
15		Mix salad	0.87	0.0088	0.0085	0.0086
16	Fruits	Apples	0.051	0.0139	0.0162	0.0162
17		**Bananas**	**<u>0.004</u>**	0.0638	0.0701	0.062
18		**Strawberries**	**<u>0.000</u>**	0.0216	0.0296	0.0323
19		**Avocado**	**<u>0.000</u>**	0.0126	0.0285	0.0302
20		**Blueberries**	**0.000**	0.0104	0.0093	0.0144
21		**Grapes**	**0.000**	0.0158	0.0084	0.0092
22	Soft drinks	Coca-Cola	0.976	0.0304	0.03	0.0299
23		Own drinks	0.517	0.0249	0.0228	0.0238
24		Schweppes	0.127	0.0108	0.0103	0.0119
25	Biscuits	**Arnott's**	**0.006**	0.0429	0.0478	0.0425
26		Own biscuits	0.158	0.0133	0.0149	0.014
27	Snacks	Smiths	0.207	0.0246	0.0254	0.023
28		Own snacks	0.832	0.0104	0.0107	0.0108
29		Doritos	0.555	0.009	0.0093	0.0096
30	Confectionery	**Cadbury**	**0.018**	0.0187	0.022	0.0193
31		Own confectionery	0.679	0.0083	0.0087	0.0082
32		Mars	0.127	0.0087	0.0097	0.0089
33	Yoghurt	Own yoghurt	0.164	0.0269	0.0294	0.0277
34		Yoplait	0.264	0.0118	0.0124	0.011
35		Vaalia	0.289	0.0096	0.0097	0.0085
36	Cereal	Kelloggs	0.126	0.015	0.0147	0.0134
37		Uncle Toby	0.052	0.0108	0.0111	0.0096
38		Sanitarium	0.304	0.0128	0.0134	0.0123

There are 12 products (*p-value* and product names are in bold) with statistically significantly different purchase rates across three phases ($p < 0.05$). Among these 12 products, 10 are from the Fruits & Vegetables category, one is from Biscuits, and one is from Confectionery. In more detail, 7 out of the 10 fresh products (their *p-values* are underlined) have significantly higher mean purchase rates in Phase 2 than one or both of the other two phases, which shows the increase in preferences towards these products during the program. For the other 3 fresh products, truss tomatoes, blueberries and grapes, the purchase rates in Phase 2 are not the largest, although the purchase rates of three phases are significantly different. We notice that the purchase rates of these 3 products are strongly negatively correlated with product prices (Pearson's correlations are −0.28, −0.59 and −0.64, respectively), so the lower purchase rates in Phase 2 may be caused by the strong seasonal effects.

There are 7 out of 21 (33.3%) fresh products with significantly higher purchase rates in Phase 2, while this ratio is 2 out of 17 (11.8%) for other products. This partially demonstrates the impact of the health program. However, the mean purchase rates cannot capture how purchase behaviour changes over time, and it is hard to distinguish the program impact from the other factors. Therefore, we investigate the distributions of different behaviour patterns and the relative increase in purchase rates.

6.3.2 Distributions of Behaviour Patterns

We investigate the pattern of individual purchase rate curve $\lambda_i(x)$, which allows us to track how customer behaviour changes during the program. For each customer, we assign one of the five patterns—increase, decrease, U-shape, inverse U-shape or stable to describe their purchase behaviour during the program. Then, we divide customers based on the four partition rules PT1–PT4, as defined in Sect. 6.2.2. We explore the distributions of the five behaviour patterns for different types of customers. The proportion of each pattern is an average over all customers in a certain group. In order to check if the distributions of patterns are correlated to the categories of products, we also compute the average proportion of each pattern across fresh products and other products.

The distributions of the behaviour patterns for various cases are shown in Table 6.3. There are four sections in the table, corresponding to the four ways of partition PT1–PT4. Each partition splits customers into two groups, and each customer group has three rows, which show their pattern distributions across all products, fresh products and non-fresh products (i.e. other products), respectively. For example, rows 1, 3 and 5 are for active web users. The most dominant pattern for each case is highlighted in bold.

Active versus inactive web users (PT1)

For these two groups, we notice that the average proportions of U-shape, inverse U-shape and stable patterns for all products are significantly different, while the proportions of increase and decrease patterns are similar.

Table 6.3 Distributions of five purchase behaviour patterns for different customer groups during the program. The most dominant pattern for each case is highlighted in bold

Partition	Products	Customer group	Increase (%)	Decrease (%)	U-shape (%)	Inverse U (%)	Stable (%)
PT1	All	Active web users	17.12	15.85	10.02	18.54	**38.47**
		Inactive web users	17.64	14.91	3.94	4.90	**58.61**
	Fresh	Active web users	15.60	14.00	14.19	18.08	**38.14**
		Inactive web users	18.91	13.52	6.25	4.59	**56.72**
	Others	Active web users	19.00	18.15	4.87	19.10	**38.89**
		Inactive web users	16.07	16.62	1.09	5.28	**60.94**
PT2	All	Active diary users	16.90	15.65	11.21	20.44	**35.80**
		Inactive diary users	17.42	15.22	5.20	8.13	**54.04**
	Fresh	Active diary users	16.06	15.33	15.35	18.51	**34.75**
		Inactive diary users	17.77	13.15	8.04	8.31	**52.73**
	Others	Active diary users	17.94	16.03	6.09	22.83	**37.11**
		Inactive diary users	16.98	17.78	1.68	7.90	**55.65**
PT3	All	2 valid weigh-ins	19.81	15.04	15.81	**36.39**	12.95
		<2 valid weigh-ins	17.17	15.42	5.73	9.05	**52.64**
	Fresh	2 valid weigh-ins	12.16	14.06	21.80	**35.60**	16.39
		<2 valid weigh-ins	17.83	13.66	8.58	8.79	**51.14**
	Others	2 valid weigh-ins	29.26	16.25	8.40	**37.37**	8.71
		<2 valid weigh-ins	16.35	17.59	2.21	9.36	**54.49**
PT4	All	≥3 kg weight loss	18.89	14.75	14.82	**43.64**	7.90
		<3 kg weight loss	19.55	14.99	16.73	**30.01**	16.09
	Fresh	≥3 kg weight loss	10.20	15.12	21.38	**42.91**	10.39
		<3 kg weight loss	13.72	13.04	21.33	**32.10**	19.82
	Others	≥3 kg weight loss	29.62	14.29	6.72	**44.55**	4.82
		<3 kg weight loss	28.44	18.49	11.74	**29.14**	12.20

Specifically, the active web users have significantly fewer stable patterns ($p < 0.05$), but more U-shape ($p < 0.001$) and inverse U-shape ($p < 0.01$) patterns than the inactive web users. This means that the purchase rates of active web users have more fluctuations during the program than of the other customers. After splitting the products into fresh and other categories, for the active web users, the amount of U-shape pattern is 14.19% for fresh products, and 4.87% for other products, with a large gap between them. For the active web users, the proportions of the other four patterns between fresh and other products are close to each other, with the gaps ranging from 0.75 to 4.15%. As for the inactive web users, the amount of U-shape pattern is 6.25% for fresh products, and 1.09% for other products. The increase in the proportion of U-shape pattern of fresh products for both active and inactive web users could be caused by the seasonal effect, as the purchase rates of fresh products may first decrease when the program started in winter and increase when the program impact became stronger than the seasonal effect.

Active versus inactive diary users (PT2)

As for PT2, the amount of inverse U-shape pattern for the active diary users is 20.44%, which is significantly higher than the inactive diary users, 8.13% ($p < 0.001$). For the active diary users, the amount of stable pattern is 35.80%, which is about 20% lower than for the inactive diary users, 54.04% ($p < 0.01$). The differences between the two groups of the other three patterns are not significant.

For the active diary users, the proportions of first four patterns for fresh products range from 15 to 18%, whereas the proportion of stable pattern is much higher, 34%. Similar to the active web users, the active diary users also have significantly more U-shape pattern for fresh products (15.35%) than other products (6.09%). Except for the U-shape pattern, the differences between fresh and other products for either customer group (e.g. comparing row 3 with 5 or comparing row 4 with 6 of the second section in Table 6.3) are less than 5%. This shows that the difference between fresh and other products for one customer group is less pronounced than the difference between the two customer groups. The possible reasons include that customers may buy different products together and shop for their family when they visit supermarket.

2 versus < 2 valid weigh-ins (PT3)

The customers with 2 valid weigh-ins have significantly more U-shape ($p < 0.05$) and inverse U-shape ($p < 0.001$) patterns, and fewer stable patterns ($p < 0.001$) than the other customers. The customers with 2 valid weigh-ins have more inverse U-shape patterns for both fresh and other products than their counterparts, which indicates that they were motivated by the health program at the start, but about 36% of them did not persist until the end of the program.

We notice that the gaps between two customer groups for this partition are larger than the previous two partitions PT1 and PT2. For example, the difference of the proportions of the stable pattern for two groups is about 40%, which doubles the difference between the active and inactive web users. This is a positive result, showing that the purchase behaviour of customers with or without 2 valid weigh-ins are more distinguishable. This means that the health program imposed stronger impact on the customers with higher levels of engagement, such as those using weight trackers.

≥3 kg versus <3 kg weight loss (PT4)

The groups formed by PT4 are: the customers with ≥3 kg or <3 kg weight loss, and both groups have two or more valid weigh-ins. The proportions of all patterns between these two groups are not significantly different. In this section, the most dominant pattern for all cases is inverse U-shape. The customers who lost ≥3 kg only have 7.9% stable pattern, but 43.64% inverse U-shape pattern. On average, they have 13% more inverse U-shape pattern than customers who lost <3 kg.

Compared to the customers with <2 valid weigh-ins in PT3, both groups in PT4 have significantly more inverse U-shape and fewer stable patterns. The results mean that the health program is more influential for customers who reported weigh-ins at least twice, but whether they lost greater or less than 3 kg makes smaller differences.

6.3.3 *Program Impact Scores*

In addition to the pattern distributions, we also use the impact score as defined in Sect. 6.2.2 to evaluate if the health program is more effective for a certain customer group. The impact score is designed to compare the relative increase in the purchase rates of different customer groups. We use partition criteria PT1–PT4 again to group all customers, and our target groups are: active web users, active diary users, the customers with 2 valid weigh-ins and the customers with $\geq$3 kg weight loss, respectively; and the other customer groups are all control groups.

For each product, the higher impact score is preferred, which means that the customers from the target group have larger increases in purchase rates during the program than other customers. For example, impact score 0.6 means 20% lift from the baseline 0.5. Therefore, we rank impact scores across all products for each partition method, and present the results in Tables 6.4 and 6.5. Each table corresponds to one partition method, and the table contains rankings, product names (the fruits and vegetables ranked top 20 are highlighted in bold) and impact scores.

When the target group is active web users (PT1, in Table 6.4-left), the scores of all products are between 0.48 and 0.62. As for the top 20 products, there are 12 fresh products, and 4 of them are fruits: grapes, strawberries, blueberries and apples. The proportion of fruits among top 20 scores is higher than other categories, given there are 6 fruits involved in the study. The proportion of fresh products appearing among top 20 is 57% (12 out of 21), while it is 47% (8 out of 17) for other products. The result shows that the active web users have slightly larger increase in the purchase rates of fresh products, especially fruits, than others.

For the active diary users (PT2, in Table 6.4-right), there are also 12 fresh products among the products with top 20 impact scores. The truss tomatoes rank first, with impact score 0.7, which is 40% lift from the baseline 0.5. The range of impact scores is between 0.43 and 0.7, which is wider than the range for the active web users. Especially, the scores for the first 7 products are all greater than 0.62, which means that the active diary users buying these products have larger increase in their purchase rates, when compared to their active web users counterparts.

For the third partition method (PT3, in Table 6.5-left), the target group is smaller in this case, because the number of customers reported 2 valid weigh-ins is much lower than the number of active web or diary users. This leads to a larger variance of impact scores, and the range of impact scores is between 0.32 and 0.95. The top 10 scores are all higher than 0.7, which equals to the highest value for PT2. Similarly, for the customers with $\geq$3 kg weight loss (PT4, in Table 6.5-right), the top 6 scores are all higher than 0.85. The number of fresh products among top 20 is 11, which is close to the other cases.

For both PT3 and PT4, the increase in purchase rates of the target groups are more distinguishable from the control groups than the previous two partition methods. The result shows that the customers with higher levels of interactions such as using the weight tracker (PT3 and PT4) are more receptive to the program than the rest.

Table 6.4 Left: the program impact scores for active web users (PT1); right: the program impact scores for active diary users (PT2). The fruits and vegetables in top 20 are highlighted in bold

	Active web users			Active diary users	
	Product	Impact		Product	Impact
1	**Grapes**	0.62	1	**Truss tomatoes**	0.7
2	Mars	0.62	2	**Spring onions**	0.68
3	**Truss tomatoes**	0.61	3	Uncle tobys	0.68
4	**Spring onions**	0.6	4	Yoplait	0.65
5	**Strawberries**	0.58	5	Mars	0.65
6	Doritos	0.57	6	Coca-Cola	0.64
7	Yoplait	0.57	7	**Grapes**	0.63
8	Uncle tobys	0.56	8	Doritos	0.62
9	**Red capsicum**	0.56	9	Own drinks	0.6
10	Coca-Cola	0.56	10	**Cucumbers**	0.59
11	**Cucumbers**	0.56	11	**Leb. cucumbers**	0.58
12	Own drinks	0.56	12	**Strawberries**	0.58
13	Schweppes	0.55	13	**Blueberries**	0.57
14	**Lettuce**	0.55	14	**Red capsicum**	0.57
15	**Blueberries**	0.55	15	**Field tomatoes**	0.57
16	Own snacks	0.54	16	Smiths	0.56
17	**Leb. cucumbers**	0.54	17	**Carrots**	0.56
18	**Mix salad**	0.53	18	**Cup mushrooms**	0.54
19	**Broccoli**	0.53	19	**Lettuce**	0.54
20	**Apples**	0.53	20	Schweppes	0.54
21	Smiths	0.53	21	Avocado	0.53
22	Own biscuits	0.52	22	Own confectionery	0.53
23	Grape tomatoes	0.52	23	Kelloggs	0.53
24	Own yoghurt	0.52	24	Cadbury	0.52
25	Cup mushrooms	0.52	25	Apples	0.52
26	Own confectionery	0.52	26	Sweet potatos	0.52
27	Carrots	0.51	27	Mix salad	0.52
28	Kelloggs	0.50	28	Broccoli	0.51
29	Arnott's	0.50	29	Vaalia	0.51
30	Sweet potatos	0.50	30	Bananas	0.50
31	Avocado	0.50	31	Own biscuits	0.50
32	Bananas	0.50	32	Grape tomatoes	0.50
33	Cadbury	0.50	33	Arnott's	0.50
34	Field tomatoes	0.49	34	Own yoghurt	0.48
35	Red onions	0.49	35	Own snacks	0.47
36	Vaalia	0.48	36	Zucchini	0.46
37	Sanitarium	0.48	37	Red onions	0.46
38	Zucchini	0.48	38	Sanitarium	0.43

Table 6.5 Left: The program impact scores for the users with 2 valid weigh-ins (PT3); right: the program impact scores for the users with ≥3 kg weight loss (PT4). The fruits and vegetables in top 20 are highlighted in bold

	2 valid weigh-ins			≥3 kg weight loss	
	Product	Impact		Product	Impact
1	Doritos	0.95	1	Yoplait	0.99
2	**Truss tomatoes**	0.8	2	**Spring onions**	0.99
3	Schweppes	0.8	3	Doritos	0.93
4	**Strawberries**	0.79	4	Schweppes	0.89
5	Yoplait	0.79	5	**Truss tomatoes**	0.88
6	**Grapes**	0.79	6	**Strawberries**	0.86
7	Own drinks	0.78	7	**Blueberries**	0.79
8	Coca-Cola	0.76	8	**Red capsicum**	0.78
9	Uncle tobys	0.76	9	**Grape tomatoes**	0.78
10	**Spring onions**	0.73	10	Uncle tobys	0.78
11	**Lettuce**	0.7	11	Own drinks	0.78
12	Own confectionery	0.7	12	**Grapes**	0.75
13	Vaalia	0.69	13	Vaalia	0.73
14	**Red capsicum**	0.68	14	**Cucumbers**	0.7
15	**Grape tomatoes**	0.68	15	**Broccoli**	0.67
16	**Cucumbers**	0.67	16	**Red onions**	0.67
17	Smiths	0.65	17	Own confectionery	0.66
18	**Red onions**	0.64	18	Kelloggs	0.66
19	**Blueberries**	0.63	19	Smiths	0.65
20	**Apples**	0.63	20	**Lettuce**	0.64
21	Broccoli	0.62	21	Bananas	0.64
22	Leb. cucumbers	0.62	22	Apples	0.61
23	Carrots	0.61	23	Mars	0.60
24	Mars	0.60	24	Zucchini	0.56
25	Cadbury	0.59	25	Coca-Cola	0.56
26	Avocado	0.58	26	Sweet potatos	0.55
27	Kelloggs	0.57	27	Carrots	0.54
28	Arnott's	0.55	28	Leb. cucumbers	0.54
29	Bananas	0.55	29	Avocado	0.52
30	Own yoghurt	0.53	30	Arnott's	0.50
31	Zucchini	0.53	31	Cadbury	0.50
32	Cup mushrooms	0.52	32	Own snacks	0.49
33	Mix salad	0.50	33	Own yoghurt	0.46
34	Sweet potatos	0.50	34	Cup mushrooms	0.45
35	Field tomatoes	0.48	35	Field tomatoes	0.45
36	Own snacks	0.43	36	Mix salad	0.43
37	Own biscuits	0.39	37	Own biscuits	0.34
38	Sanitarium	0.32	38	Sanitarium	0.11

We find that the products from the other categories, such as Uncle Tobys from Cereal, Yoplait from Yoghurt and Doritos from Snacks, also have high scores for all four partitions. Although the health program did not have discounts on them, this should not be interpreted as negative results. The main reason is that customers may buy them together with fresh products when they visit supermarket [8], so we could see their purchase rates change with fresh products simultaneously. This has also been discussed in Sect. 6.3.2, as we find the differences of pattern distributions between fresh and other products for the active web users are small.

It is also worth noting that grapes, truss tomatoes and blueberries have significantly lower mean purchase rates during the program (Phase 2) than the other phases as discussed in Sect. 6.3.1, but their impact scores appear in top 20 for all four partitions. Especially, for the customers with ≥3 kg weight loss, the scores are 0.88, 0.79 and 0.75 for these three products. The high scores for these products show that the target customers have larger increases in the purchase rates than other customers, even though the overall purchase rates are lower at that time of the year. As discussed, the lower absolute purchase rates could be caused by seasonality or other factors, while the higher increases in the purchase rates indicate that the target customers are more receptive to the health program than other customers. This shows the advantage of using the impact score metric, when it is hard to isolate the impact of the health program from the other factors.

6.4 Summary

In this chapter, we propose a method for evaluating the impact of a web-based health program by tracking individual behaviour changes and conducting a fine-grained analysis across the groups of customers and products. To better analyse the impact of the health program, we use both purchase data and usage logs from the associated website of the health program. Our method allows uncovering the hidden dependencies and impact, which may be overlooked by the traditional analysis methods without behavioural tracking.

Since the individual purchase records are sparse and noisy, we construct accurate individual purchase rate curves based on a mixture of segment-level NHPP models, which facilitates the evaluation of the program impact. We also design four criteria for partitioning the customers and compare the program impact on the customers with different levels of interactions with the program website. The key findings from our case study on the supermarket health program are:

- For 7 fresh products and 2 other products, the average purchase rates during the program are significantly higher than before or after the program.
- The active web users, active diary users, the customers with 2 valid weigh-ins and the customers with ≥3 kg weight loss have been motivated to a greater extent to change their purchase behaviour. They have significantly fewer stable patterns, and more inverse U-shape patterns than the others. Based on the impact scores, they

have larger increases in their purchase rates for some fresh products, especially fruits.

- Among four partition methods, when the criterion considers higher levels of interactions such as using the weight tracker (PT3 and PT4), the program impact is more pronounced on the target groups than the rest. The customers who reported at least 2 valid weigh-ins and who lost more than 3 kg achieved larger behaviour changes than others.

The results have shown the important role of the program website, which provides interactive and personalised tools to encourage behaviour changes. The interactions with the participants (e.g. for food diary and tracking weight loss) can be further improved by using mobile applications. Our method can be used by the designers of health programs to understand the behaviour of their participants, increase the participant's engagement and target different types of participants.

References

1. Ball K, McNaughton SA, Le HN, Gold L, Ni Mhurchu C, Abbott G, Pollard C, Crawford D (2015) Influence of price discounts and skill-building strategies on purchase and consumption of healthy food and beverages: outcomes of the supermarket healthy eating for life randomized controlled trial. Am J Clin Nutr 101(5):1055–1064
2. Brindal E, Hendrie G, Freyne J, Coombe M, Berkovsky S, Noakes M (2013) Design and pilot results of a mobile phone weight-loss application for women starting a meal replacement programme. J Telemed Telecare 19(3):166–174
3. Dickinson WP, Glasgow RE, Fisher L, Dickinson LM, Christensen SM, Estabrooks PA, Miller BF (2013) Use of a website to accomplish health behavior change: if you build it, will they come? And will it work if they do? J Am Board Fam Med 26(2):168–176
4. Wieland LS, Falzon L, Sciamanna CN, Trudeau KJ, Brodney S, Schwartz JE, Davidson KW (2012) Interactive computer-based interventions for weight loss or weight maintenance in overweight or obese people. Cochrane Database Syst Rev 8(8):
5. Brindal E, Freyne J, Saunders I, Berkovsky S, Smith G, Noakes M (2012) Features predicting weight loss in overweight or obese participants in a web-based intervention: randomized trial. J Med Internet Res 14(6):e173
6. Turner JR, Thayer J (2001) Introduction to analysis of variance: design, analysis & interpretation. Sage
7. Manning C, Raghavan P, Schütze H (2008) Introduction to information retrieval. Cambridge University Press, Cambridge
8. Larson JS, Bradlow ET, Fader PS (2005) An exploratory look at supermarket shopping paths. Int J Res Mark 22(4):395–414

Chapter 7
Tracking the Evolution of Customer Segmentations

7.1 Background

While traditional customer segmentation approaches are based on the demographic information, recent approaches focus on identifying different types of customers in terms of their purchase behaviour. For example, mixture models [1] can be utilised to discover latent groups of customer behaviour. In this way, the purchase behaviour of an individual customer can be interpreted as a mixture of different prototypes of behaviour weighted by the membership in those groups.

However, the mixture modelling with static membership is not suitable for behaviour modelling in dynamic scenarios, e.g. for tracking a customer's purchase behaviour shifting across groups. For example, a customer Alice who bought potato chips regularly, may buy less of them after she embarked on a diet. Considering the group level, Alice may shift from Group 1 comprising customers who prefer snacks, to Group 2 comprising customers who prefer fruits. If all the other customers remain invariant, it is easy to track Alice's change by simply updating her group membership. However, other customers may also change their behaviour at the same time, which leads to shifting the behaviour of the entire group (prototype). Alice may still have a large or even unchanged membership of Group 1, if many other customers from Group 1 also shift to Group 2. This may provide misleading results to business management, who desire to know how the behaviour of each customer changes across groups over time and which customers are more receptive to promotions.

A better way for dealing with the "group-level behaviour shifting" problem is by adopting a split-merge approach. On one hand, a group can split into multiple smaller groups, when different types of behaviour emerge in the group; on the other hand, the customers from two or more groups can merge into a large group, when they are affected by common factors, such as changes of seasons and price—which are of interest for the stakeholders. Due to the aforementioned limitations of using mixture models to address certain demands of the stakeholders, we propose a Bayesian nonparametric Customer Segmentation Model, named FC-CSM, which can track varying customer groups via a Fragmentation-Coagulation Process (FCP).

L. Luo, *Temporal Modelling of Customer Behaviour*,
Springer Theses, https://doi.org/10.1007/978-3-030-18289-2_7

The FCP is a random partition process, which can describe the split and merge of groups over a period of time [2]. At each time interval, the model conducts fragmentation and coagulation operations on all the existing groups in the partition. Therefore, one group can split into smaller groups in the fragmentation step, when the members start to have divergent purchase behaviour. On the contrary, several groups can merge in the coagulation step, when customers from different groups exhibit similar behaviour. Another advantage of FCP is that, by configuring the split and merge rates, it can be employed to analyse individuals in a minor but interesting group of customers receptive to a promotion or with abnormal behaviour, which is often desired by the stakeholders.

Problem Formulation and Organisation

Given a transaction dataset, we obtain customer purchase behaviour matrix $\boldsymbol{X}^{U \times T}$ of U customers during $(0, T]$, where x_{it} is the behaviour data for customer i during $(t-1, t]$. Matrix $\boldsymbol{X}$ can represent different types of purchase behaviour. For example, x_{it} can be a binary indicator of whether customer i purchased a product during $(t-1, t]$, or it can represent the number of purchase events in week t (time unit is one week), or it can reflect a preference level for a product.

Our research tasks are to model customer behaviour and understand their behaviour changes based on purchase records $\boldsymbol{X}$. We first segment customers into groups based on the similarity of their behaviour. Then we track the evolution of customer groups, such as how the groups split and merge over time. Based on the individual customer's memberships of different groups, we can capture how individual customers shift across groups and change their purchase behaviour over time. The proposed model FC-CSM facilitates flexible customer purchase behaviour segmentation, and it can answer questions such as how many types of customers exist, when and how customer groups split and merge, how long a group lasts for, who are the customers in each group, and how receptive they are to promotions.

This section introduces the contributions and related work of this chapter. The rest of this chapter is organised as follows. Section 7.2 firstly explains the preliminary knowledge of this chapter, including Chinese Restaurant Process, FCP, maximum a posteriori and conjugate pairs. Then, the section describes the construction of customer partition sequences and the inference procedure of FC-CSM in detail. Section 7.3 introduces our case study on the supermarket purchase dataset and analyses the results of dynamic customer segmentations. Section 7.4 summarises our findings and contributions of this chapter.

Contributions

We conduct a case study on the supermarket transaction dataset. In the case study, we construct FC-CSM to explore the purchase behaviour of various popular products and answer to the above practical questions. Our contributions are listed as follows:

- We evaluate the fitness of FC-CSM and demonstrate that FC-CSM has achieved a higher accuracy than the mixture of Poisson processes.
- We examine the impact of promotions on purchase behaviour and find that customers buying fresh products, soft drinks and confectionery are more receptive to the promotions.

- We illustrate how the customer segmentations for different products evolve over time, and track how individual customers shift across groups.

The results of our work can be used by stakeholders to discover the patterns of purchase behaviour changes that are worth noting, analyse responses of various customers to promotions, and identify target customers for promotions.

Related Work

We have found that modelling all customers together may overlook their heterogeneity, while building individual models can overfit the sparse and noisy purchase events of an individual customer. The customer segmentation based on behaviour via mixture models can overcome the above disadvantages. The mixture models can discover different types of purchase behaviour to support group-level analysis [1]. The mixture models can capture behaviour changes by a dynamic temporal component in the group-level model [3–5]. The dynamic latent variable models can also be adapted to group customers. For example, the dynamic topic model in [6] and the sequential latent Dirichlet allocation [7], which are both based on the popular topic modelling technique—Latent Dirichlet Allocation (LDA).

To explore how customer groups change over time, previous work examined the groups at different time points and detected changes based on rules. Spiliopoilou et al. [8] proposed a framework MONIC to monitor cluster transition. They re-clustered members at each time point and utilised a set of rules to match clusters from different time points. They could detect the external and internal transitions of clusters, such as the survival, split, expansion and disappearance of clusters. Another option is to conduct group adaptation at each time step, which can avoid matching groups afterwards in a heuristic way. The group adaptation can also consider the temporal dependency of the grouping structure from neighbouring time steps in the model, instead of generating a set of new groups at each time point. Xing and Sohn [9] proposed a Hidden Markov Model (HMM), which matched the possible states of an HMM to all groups of the partition. Since an HMM needs to label all states, the model suffers from the label switching problem, where each permutation of labels generates a mode in the posterior distribution, which may lead to non-identifiability [10].

The random partition process FCP can describe the evolution of partitions using the fragmentation and coagulation operations [2]. Moreover, as a group of FCP is determined by its members rather than an explicit group label, the FCP can avoid the label switching problem. The idea of FCP has been applied in different fields such as financial markets [11], evolving networks [12], and genetic variation [13, 14]. Our proposed model fills the gap of existing work on tracking the evolution of customer groups and how individual customers shift across groups via a split-merge approach.

7.2 Methodology

In this section, we introduce the preliminary concepts of FC-CSM, describe the construction of FC-CSM from a generative perspective, and explain how to use the Gibbs sampler to infer the parameters of our model.

7.2.1 Preliminaries

Chinese Restaurant Process and Fragmentation-Coagulation Process

At time t, the set of all customers A can be divided into n_t non-empty and non-overlapping groups, whose union is A. These groups form a partition π_t of A, where each group contains customers with similar behaviour at t.

The Chinese Restaurant Process (CRP) [15] is a nonparametric random partition model, which can be interpreted by the analogy of customers choosing tables in a Chinese restaurant. Assuming the restaurant has an infinite number of tables with infinite capacity, the first customer can choose any table, and each of the following customer can either sit with previous customers or start a new unoccupied table. For each customer, the probability of choosing an existing table is proportional to the number of customers at that table, while the probability of choosing a new table is proportional to the number of tables which have been occupied. After all customers find their tables, it forms a partition of these customers, and each table corresponds to a customer group in the partition. CRP can avoid predefining the number of groups in the partition, which allows for a great flexibility of the model.

Formally, as for $CRP(A, \rho, \theta)$, ρ ($\rho > 0$) is the *strength* parameter, which controls the number of groups in the partition, and θ is the *discount* parameter for the process. Followed by $CRP(A, \rho, \theta)$, the probability of choosing group k for customer i given the allocation of previous customers is as follows:

$$P(z_i = k|z_1, z_2, \ldots, z_{i-1}) = \begin{cases} \dfrac{n\theta + \rho}{i - 1 + \rho} & \text{if group } k \text{ is a new group,} \\ \dfrac{|S_k| - \theta}{i - 1 + \rho} & \text{if group } k \text{ is an existing group.} \end{cases} \tag{7.1}$$

where n is the number of existing groups formed by previous $i - 1$ customers, set S_k is the selected group for the customer, with group index k; $|S_k|$ is the number of customers in S_k. Equation 7.1 shows that the probability of starting a new group is higher when there are more existing groups, and the probability of selecting an existing group is higher for the groups with more customers.

The random partition process FCP is defined based on CRP. In more detail, the marginal distribution of the partition at a certain time for FCP is a CRP. FCP conducts the fragmentation and coagulation operations on all groups of the partition, so that the groups can split and merge to form a new partition at the next time interval. In the fragmentation step, each group either remains unchanged or splits into smaller groups. On the contrary, in the coagulation step, each group either remains unchanged or merge with other groups. The marginal distribution of the partitions for an FCP remains unchanged after a series of fragmentation and coagulation operations. As outputs, we can get a sequence of partitions $\{\pi_t\}$ ($t \in \{1, \ldots, T\}$), which allows us to know who are in the same group at any time and how customer groups evolve over time.

Maximum a Posteriori and Conjugate Pairs

Maximum *a posteriori* (MAP) is to estimate parameter θ by maximising the posterior probability given observation data $\boldsymbol{x}$. MAP includes a priori belief $p(\theta)$ of the parameter, which can carry some prior knowledge about the model and avoid overfitting the observation [16]. Based on the Bayes' rule, the optimisation objective of MAP is:

$$\begin{aligned}\hat{\theta} &= \underset{\theta}{argmax}\ p(\theta|\boldsymbol{x}) \\ &= \underset{\theta}{argmax}\ \frac{p(\boldsymbol{x}|\theta)p(\theta)}{p(\boldsymbol{x})}\end{aligned} \tag{7.2}$$

The parameter θ is considered as a random variable, and it has hyperparameter α for its distribution, which means $p(\theta) := p(\theta|\alpha)$.

To facilitate the model inference, the conjugate prior for the observation data can be used, so that the prior distribution $p(\theta)$ and the posterior distribution $p(\theta|\boldsymbol{x})$ have the same form but different parameters [17]. The conjugate pair can not only simplify the computation of the inference, but makes it easier to interpret the hyperparameters. For example, let us consider the conjugate pair Beta and Bernoulli distributions. Given a prior distribution $Beta(\alpha, \beta)$ for Bernoulli observation $\boldsymbol{x}$, the posterior distribution follows $Beta(\alpha + n_1, \beta + n_0)$, where n_1 is the number of 1's in $\boldsymbol{x}$ and n_0 is the number of 0's in $\boldsymbol{x}$. The hyperparameters α and β can be interpreted as pseudo-counts of 1 and 0 before observing $\boldsymbol{x}$.

In our model, the conjugate pair is Gamma and Poisson distributions, as Poisson distribution is often used to describe the number of random points in a temporal or spatial space [18]. For $Poisson(\lambda)$, the prior of λ is $Gamma(\alpha, \beta)$, which has the probability density function as:

$$p(\lambda|\alpha, \beta) = \frac{\lambda^{\alpha-1}e^{-\lambda/\beta}}{\beta^{\alpha}\Gamma(\alpha)} \tag{7.3}$$

where α ($\alpha > 0$) is the shape parameter, β ($\beta > 0$) is the scale parameter, and $\Gamma(\alpha) = \int_0^{+\infty} y^{\alpha-1}e^{-y}dy$.

Given observation $\boldsymbol{x}$, the posterior distribution of λ is:

$$\begin{aligned}p(\lambda|\boldsymbol{x}, \alpha, \beta) &= \frac{\lambda^{\alpha'-1}e^{-\lambda/\beta'}}{\beta'^{\alpha'}\Gamma(\alpha')} \\ \alpha' = \alpha + \sum_{i=1}^{n} x_i, \quad \beta' &= \frac{\beta}{1 + n\beta}\end{aligned} \tag{7.4}$$

For $\alpha \geq 1$, $Gamma(\alpha, \beta)$ has a single mode (i.e. maximal value) at $(\alpha - 1)\beta$, so that the MAP estimation of λ is:

$$\hat{\lambda} = (\alpha' - 1)\beta' = \frac{(\alpha + \sum_{i=1}^{n} x_i - 1)}{\beta^{-1} + n} \tag{7.5}$$

Equations 7.4 and 7.5 show the advantage of using conjugate pairs in the inference. Besides the Gamma-Poisson distribution, the other frequently used conjugate pairs include the Beta-Binomial distribution and Dirichlet-Multinomial distribution.

7.2.2 Construct Sequence of Customer Partitions

This section will describe the construction of our model FC-CSM. The generative process of FC-CSM is defined as follows:

$$\begin{aligned} \pi_1 &\sim CRP(A, \rho, 0) \\ \pi_{t+1}|\pi_t &\sim FCP(\pi_t, \rho, \delta) \\ x_{it}|\pi_t(i) = k &\sim Poisson(\lambda_{\pi_t(i)}) \\ \lambda_k &\sim Gamma(\alpha, \beta) \end{aligned} \tag{7.6}$$

Sample Sequence of Partitions

We adopt the CRP to describe the partitioning of all customers, that is, customer segmentation. From the perspective of the generative model, at $t = 1$, we draw π_1 for customer set A from $CRP(A, \rho, 0)$. Then we obtain the following partition π_{t+1} based on $FCP(\pi_t, \rho, \delta)$, where δ ($\delta \in [0, 1)$) controls the *temporal dependence* between π_t and π_{t+1}. When $\delta = 0$, the partition fully depends on the previous one, which means that the partition remains the same from t to $t + 1$; while $\delta \to 1$ means that π_t and π_{t+1} are independent. Any group in π_t can split into smaller groups and merge with the other groups to generate π_{t+1}.

Given π_1, the transition from π_t to π_{t+1} at $t \in \{1, \ldots, T - 1\}$ can be completed in two sub-steps: the first step *splits* all groups in π_t into smaller groups and produces an intermediate partition π_t'; while the second step *merges* groups in π_t' to produce π_{t+1}. These two steps correspond to the fragmentation (split) and coagulation (merge) operations, respectively.

Specifically, the FCP transition rules are:

(1) Each group in π_t' either exists in π_t or is a subgroup of another group in π_t.
(2) Each group in π_{t+1} either exists in π_t' or is a combination of other groups in π_t'.

For example, Fig. 7.1 illustrates the sequence of partitions for 10 customers from t to $t + 2$ with different scenarios. From π_t to π_t', the top group with 7 customers splits into two smaller groups with 2 and 5 customers, while the bottom group remains unchanged, which is equivalent to splitting to an empty group and the group with the same customers as the original group. From π_t' to π_{t+1}, the top group remains unchanged, while the other two groups merge into one group. From π_{t+1} to π_{t+1}', it shows that one group can split into multiple groups, instead of just two.

FCP assumes that the partition sequence $\{\pi_t\}$ ($t \in \{1, \ldots, T\}$) satisfies Markov property. It means that given π_t, π_{t+1} is conditionally independent on partitions at

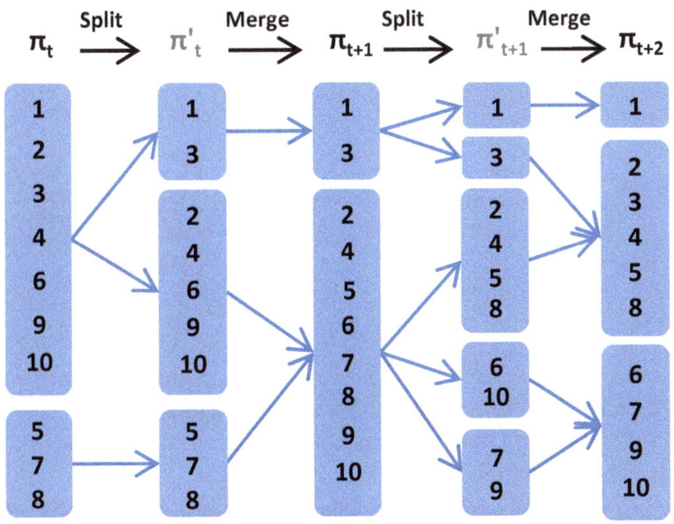

Fig. 7.1 Sequence of partitions for 10 customers from t to $t + 2$. Each blue block represents one group

the other time intervals. The fragmentation and coagulation steps do not affect the marginal distribution, so $P(\pi_t)$ at any t is $CRP(A, \rho, 0)$, and the marginal distribution of intermediate partition $P(\pi'_t)$ is $CRP(A, \rho, \delta)$.

Sample Behaviour Data

The partition of customers is dictated by the behaviour metrics determined by the requirements of analysis. For instance, if we desire to distinguish customers by the number of times they buy a product fortnightly, the purchase behaviour data x_{it} stores the number of purchase events in every two weeks. The Poisson distribution can be used to model the number of events in each period of time. Specifically, the parameter λ of the Poisson distribution can be interpreted as the expected number of events per unit of time. Therefore, we draw x_{it} from a Poisson distribution with intensity $\lambda_{\pi_t(i)}$ for customer i in group $\pi_t(i)$. In addition, we use the conjugate prior, $Gamma(\alpha, \beta)$ for the Poisson distribution.

The main advantage of our model is that we can customise the grouping of customers according to the analysis requirements. As mentioned above, the customer behaviour can be defined in different ways, by storing different data in $\boldsymbol{X}$. The changes would only affect the sampling of behaviour data, which means that we could replace Gamma-Poisson by other distributions, but the sampling of partition sequence remains unchanged. For example, if the behaviour data x_{it} is the binary purchase indicator, x_{it} could be sampled from a Bernoulli distribution with a Beta prior; if x_{it} is a preference rating, x_{it} could be sampled from a Multinomial distribution with a Dirichlet prior.

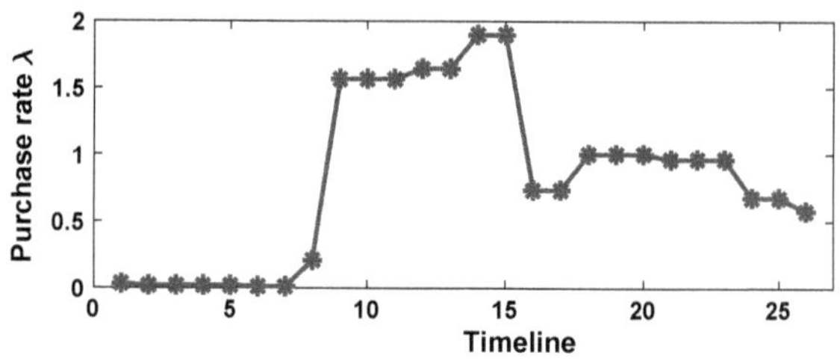

Fig. 7.2 Individual purchase rate curve (time unit: two weeks)

Individual Purchase Rate

For each customer i, we can trace how they shift across groups and their purchase rate at any time based on the sequence of their group membership $\{\pi_t(i)\}$ $(t \in \{1, \ldots, T\})$, where $\pi_t(i)$ is the group containing customer i in π_t. The purchase rate $\lambda_i(t)$ is a discrete step function defined as:

$$\lambda_i(t) = \lambda_{\pi_t(i)} \qquad t \in \{1, \ldots, T\} \tag{7.7}$$

For example, the individual purchase rate curve is illustrated in Fig. 7.2, where '$*$' denotes the value of λ, and the solid line between $\lambda_{\pi_t(i)}$ and $\lambda_{\pi_{t+1}(i)}$ is the linear interpolation of them.

Therefore, we can obtain the purchase rate curve $\lambda_i(t)$ of customer i, which captures the changes of their purchase rate, besides the partition sequence $\{\pi_t\}$ $(t \in \{1, \ldots, T\})$.

7.2.3 *Estimate Parameters of FC-CSM*

In FC-CSM (as defined in Eq. 7.6), CRP has *exchangeable* and *projective* properties, which means that:

(1) The order of allocating customers to groups does not affect the marginal distribution $P(\pi)$.
(2) The projection of π on a subset A' of the complete set A follows $CRP(A', \rho, 0)$.

Since FCP is a Markov chain with a series of split and merge operations, it is also exchangeable and projective [13]. These two properties allow us to infer the group allocation for customer i based on the current partition of the other customers, assuming customer i is the last customer who needs to be mapped to a group. Then we can update the sequence of group allocations for customer i, complying with the split and merge rules. In one sampling round, due to the exchangeability, we go through all customers to update their group allocations from $t = 1$ to $t = T$, as if this customer is the last one to join the existing customer groups.

The notation used in the description is summarised as follows: π_t^{-i} represents the projection of π_t on $A \setminus \{i\}$, which refers to set A excluding customer i; z_t is the latent group index for customer i at t; set S_k is the selected group for the customer, with group index k; $|S_k|$ is the number of customers in S_k; and $|\pi_t|$ is the number of groups in π_t. The variables with primes ($'$) are for the intermediate partition π'_t.

We adopted the forward-backward (F-B) algorithm [19] to infer the partition sequences for all customers. The F-B algorithm is a commonly used inference framework for dynamic latent variable models to compute the posterior marginals of all latent variables given observations. Basically, our F-B algorithm includes a forward phase for sampling (Eqs. 7.8–7.10) and a backward phase for smoothing (Eqs. 7.14 and 7.15).

Firstly, the posterior distributions used to sample the group allocation for customer i at $t = 1$ is represented as:

$$\begin{aligned} &P(z_1 = k|\boldsymbol{x}_i, \{\pi_\tau^{\neg i}\}_1^T, \{\pi_\tau'^{\neg i}\}_1^{T-1}) \\ \propto\ &\underbrace{P(z_1 = k|\pi_1^{\neg i})}_{\text{Equation 7.11}}\ \underbrace{P(x_{i1}|z_1 = k)}_{\text{Equation 7.18}}\ \underbrace{P(\{x_{i\tau}\}_2^T|z_1 = k, \{\pi_\tau^{\neg i}\}_1^T, \{\pi_\tau'^{\neg i}\}_1^{T-1})}_{\text{Equation 7.15}} \end{aligned} \tag{7.8}$$

where $\{\pi_\tau^{\neg i}\}_1^T$ refers to the sequence of partitions projected on $A \setminus \{i\}$ from $\tau = 1$ to $\tau = T$. In the second line, the first term is the *conditional probability* of group allocation given the partition of the other customers at time t; the second term is the *likelihood* of behaviour data x_{i1} if we select group k; and the last term is the conditional probability of all observations after the current t given the partition sequences $\{\pi_\tau^{\neg i}\}_t^T$ and $\{\pi_\tau'^{\neg i}\}_t^{T-1}$, which is also called *message*. We will explain each term in detail in its corresponding equation as noted in the under brace.

After the initialisation, we first sample k' for the *split step* at t ($t \geq 1$) based on the following posterior distribution:

$$\begin{aligned} &P(z'_t = k'|z_t = k, \boldsymbol{x}_i, \{\pi_\tau^{\neg i}\}_1^T, \{\pi_\tau'^{\neg i}\}_1^{T-1}) \\ \propto\ &\underbrace{P(z'_t = k'|z_t = k, \pi_t^{\neg i}, \pi_t'^{\neg i})}_{\text{Equation 7.12}}\ \underbrace{P(\{x_{i\tau}\}_{t+1}^T|z'_t = k', \{\pi_\tau^{\neg i}\}_t^T, \{\pi_\tau'^{\neg i}\}_t^{T-1})}_{\text{Equation 7.14}} \end{aligned} \tag{7.9}$$

After that, we sample l for the *merge step* based on the following posterior distribution:

$$\begin{aligned} &P(z_{t+1} = l|z'_t = k', \boldsymbol{x}_i, \{\pi_\tau^{\neg i}\}_1^T, \{\pi_\tau'^{\neg i}\}_1^{T-1}) \\ \propto\ &\underbrace{P(z_{t+1} = l|z'_t = k', \pi_{t+1}^{\neg i}, \pi_t'^{\neg i})}_{\text{Equation 7.13}}\ \underbrace{P(x_{i(t+1)}|z_{t+1} = l)}_{\text{Equation 7.18}} \\ &\times \underbrace{P(\{x_{i\tau}\}_{t+2}^T|z_{t+1} = l, \{\pi_\tau^{\neg i}\}_{t+1}^T, \{\pi_\tau'^{\neg i}\}_{t+1}^{T-1})}_{\text{Equation 7.15}} \end{aligned} \tag{7.10}$$

Each component in the posterior distributions (Eqs. 7.8–7.10) is introduced below. The conditional probability of allocating a group for a behaviour sequence $\boldsymbol{x}_i$ at $t = 1$ according to FCP is defined as:

$$P(z_1 = k|\pi_1^{\neg i}) = \begin{cases} |S_k|/(|A|-1+\rho) & \text{if } S_k \in \pi_t^{\neg i} \\ \rho/(|A|-1+\rho) & \text{if } S_k = \phi \end{cases} \tag{7.11}$$

where the condition $S_k \in \pi_t^{\neg i}$ in the first case means that the selected group S_k for customer i is an existing group in $\pi_t^{\neg i}$. The second case $S_k = \phi$ means that customer i will start a new group. Therefore, the possible space for S_k is $\phi \cup \pi_t^{\neg i}$.

The conditional probabilities of allocating customer i for the split and merge steps at t ($t \geq 1$) are:

$$P(z_t' = k'|z_t = k, \pi_t^{\neg i}, \pi_t'^{\neg i}) = \begin{cases} 1 & \text{if } S_k = S_{k'} = \phi \\ \delta|F_t(S_k)|/|S_k| & \text{if } S_k \in \pi_t^{\neg i}, S_{k'} = \phi \\ (|S_{k'}| - \delta)/|S_k| & \text{if } S_k \in \pi_t^{\neg i}, S_{k'} \in F_t(S_k) \\ 0 & \text{otherwise} \end{cases} \tag{7.12}$$

$$P(z_{t+1} = l|z_t' = k', \pi_{t+1}^{\neg i}, \pi_t'^{\neg i}) = \begin{cases} \rho/(\rho + \delta|\pi_t'^{\neg i}|) & \text{if } S_l = S_{k'} = \phi \\ \delta|C_t(S_l)|/(\rho + \delta|\pi_t'^{\neg i}|) & \text{if } S_l \in \pi_{t+1}^{\neg i}, S_{k'} = \phi \\ 1 & \text{if } S_l \in \pi_{t+1}^{\neg i}, S_{k'} \in C_t(S_l) \\ 0 & \text{otherwise} \end{cases} \tag{7.13}$$

Specifically, $F_t(S_k)$ refers to the set of groups in $\pi_t'^{\neg i}$ which are split from S_k; it is formally defined as $F_t(S_k) = \{B|B \in \pi_t'^{\neg i}, B \subseteq S_k, B \neq \phi\}$. Similarly, $C_t(S_l)$ refers to the set of groups in $\pi_t'^{\neg i}$ which are merged into S_l; it is defined as $C_t(S_l) = \{B|B \in \pi_t'^{\neg i}, B \subseteq S_l, B \neq \phi\}$.

The messages for the split and merge steps are denoted as mf and mc:

$$mf_t(k') = P(\{x_{i\tau}\}_{t+1}^T | z_t' = k', \{\pi_\tau^{\neg i}\}_t^T, \{\pi_\tau'^{\neg i}\}_t^{T-1}) \tag{7.14}$$

$$mc_t(k) = P(\{x_{i\tau}\}_{t+1}^T | z_t = k, \{\pi_\tau^{\neg i}\}_t^T, \{\pi_\tau'^{\neg i}\}_t^{T-1}) \tag{7.15}$$

Given $mc_T(k) = 1$ for $t = T$, the messages at other time intervals can be computed in a backward manner recursively according to the following equations:

$$mf_t(k') = \sum_{S_l \in \pi_{t+1}^{\neg i} \cup \phi} \Big(mc_{t+1}(l) \underbrace{P(z_{t+1} = l|z_t' = k', \pi_{t+1}^{\neg i}, \pi_t'^{\neg i})}_{\text{Equation 7.13}} \tag{7.16}$$

$$\times \underbrace{P(x_{i(t+1)} = |z_{t+1} = l)}_{\text{Equation 7.18}} \Big)$$

$$mc_t(k) = \sum_{S_{k'} \in \pi_t'^{\neg i} \cup \phi} mf_t(k') \underbrace{P(z_t' = k'|z_t = k, \pi_t^{\neg i}, \pi_t'^{\neg i})}_{\text{Equation 7.12}} \tag{7.17}$$

It means that $mf_t(k')$ is computed based on $mc_{t+1}(l)$ for all groups in $\pi_{t+1}^{\neg i}$ and an empty group. Then $mc_t(k)$ is computed based on $mf_t(k')$ for all groups in $\pi_t'^{\neg i}$ and an empty group. These two equations are used in turn until we get all messages for customer i.

The likelihood term $P(x_{it}|z_t = k)$ in the posterior distributions and messages is determined by the selected distribution for the observation x_{it}. Assuming x_{it} has a Poisson distribution, the likelihood of x_{it} in group k is defined as:

$$P(x_{it}|z_t = k) = \frac{\lambda_k^{x_{it}} e^{-\lambda_k}}{x_{it}!} \tag{7.18}$$

Based on the partition sequence $\{\pi_t\}$ ($t \in \{1, \ldots, T\}$), the intensity parameter of the Gamma-Poisson distribution for group k estimated by MAP is:

$$\lambda_k = \begin{cases} \dfrac{\sum_{j \in S_k} x_{jt} + \alpha - 1}{|S_k| + \beta^{-1}} & \text{if } S_k \in \pi_t^{\neg i} \\ \dfrac{\sum_{j \in A \setminus \{i\}} x_{jt} + \alpha - 1}{|A| - 1 + \beta^{-1}} & \text{if } S_k = \phi \end{cases} \tag{7.19}$$

Table 7.1 The inference procedure of FC-CSM

Input	Behaviour matrix $\boldsymbol{X}^{U \times T}$, number of sampling iterations
Output	partition sequence $\{\pi_T\}_1^T$, purchase intensity $\boldsymbol{\lambda}$

```
for iteration = 1:number_of_sampling_iterations
  for i = 1 : U
    reset messages mf and mc
    reset purchase intensity λ
    % compute messages
    for t = T : −1 : 1
      compute mf_t based on Eq. 7.16
      compute mc_t based on Eq. 7.17
    endfor
    % sample
    for t = 1 : T − 1
      if t == 1
        sample based on Eq. 7.8
        update π_1(i)
      endif
      % fragmentation step
      sample based on Eq. 7.9
      update π'_t(i)
      % coagulation step
      sample based on Eq. 7.10
      update π_{t+1}(i)
    endfor
  endfor % for one customer
endfor % for one sampling iteration
compute λ based on all customers
```

This equation derives from the basic form in Eq. 7.5. When S_k is an existing group in $\pi_t^{\neg i}$, we compute λ_k based on the other customers in S_k. Otherwise, when S_k is a new group without any customers, we compute the default λ_k based on all the customers in $A \setminus \{i\}$.

As discussed in Sect. 7.2.2, this model can be customised to handle various types of behaviour data, and the only changes refer to the third and fourth lines of the model in Eq. 7.6. In terms of inference, changing the distribution in those two lines would only affect the computation of the likelihood terms in posterior distributions.

The inference procedure of FC-CSM is illustrated in Table 7.1. In a sampling round, for each customer, we first update the sequence of messages based on the current partition sequence of the other customers (refer to lines 6–9 in Table 7.1). Then we sample new group allocations for the current customer based on Eqs. 7.8–7.10 (refer to lines 11–22). After updating the group allocations of all customers, we start the next sampling round and repeat this process until convergence. The output is the partition sequence of all customers and purchase intensity of each group.

The theoretical runtime of a sampling round is $O(UKT)$, which increases linearly with the number of customers U and the average number of groups K in each partition at time t. As multiple products can be paralleled, the model has a good scalability.

FC-CSM offers several notable advantages:

(1) It can track the *evolution of customer groups* and find the size, duration, ancestors and descendants of each customer group.
(2) It can capture the *trajectories of customers* shifting across groups and the variations of individual purchase rates.
(3) It is *customisable* and can analyse various types of behaviour data.

7.3 Case Study

Our case study is based on 38 popular products selected according to the number of customers who bought these products at least 10 times during the observation period. There are 21 fresh products including fruits and vegetables and 17 other products such as soft drinks and biscuits. The names and categories of all products are listed in Table 7.2. This is the same set of products as that have been used in Chap. 6.

The supermarket stakeholders desire to construct accurate models to understand customer behaviour, including what types of customers they have and whether supermarket promotions are effective. They also want to track the dynamics of purchase behaviour such as how the customer groups evolve over time and how individual customers shift across groups. Therefore, we conduct the case study aiming to:

(1) Evaluate the fitness of our model and compare it with two mixture models—mixture of Homogeneous Poisson Process (HomoPP) and mixture of Non-Homogeneous Poisson Process (NHPP) [3];
(2) Explore the impact of promotions on customer purchase behaviour for different products;
(3) Demonstrate the trajectories of customer partitions and how individual customers shift across groups.

Table 7.2 Categories and names of all 38 products involved in the case study

ID	Category	Product	ID	Category	Product
1	Vegetables	Carrots	20	Fruits	Blueberries
2		Mushrooms	21		Grapes
3		Red Onions	22	Soft Drinks	Coca-Cola
4		Sweet Potatoes	23		Own brand
5		Red Capsicum	24		Schweppes
6		Broccoli	25	Biscuits	Arnott's
7		Cucumbers	26		Own brand
8		Zucchini	27	Snacks	Smiths
9		Field Tomatoes	28		Own brand
10		Lettuce	29		Doritos
11		Truss Tomatoes	30	Confectionery	Cadbury
12		Leb. Cucumbers	31		Own brand
13		Grape Tomatoes	32		Mars
14		Spring Onions	33	Yoghurt	Own brand
15		Mix Salad	34		Yoplait
16	Fruits	Apples	35		Vaalia
17		Bananas	36	Cereal	Kelloggs
18		Strawberries	37		Uncle Toby
19		Avocado	38		Sanitarium

The parameters of our model are configured as: $\rho = 0.2$, $\delta = 0.1$, $\alpha = 2$, $\beta = 0.5$, and the unit of time is 14-day. We conduct 500 iterations of Gibbs sampling for each product to obtain the partition sequence. Specifically, we set $\alpha = 2$, $\beta = 0.5$ empirically based on the distribution of a validation set. We count number of purchases per two weeks of all customers, and find that the number of purchases has a Poisson distribution with mode at $x = 0$. It means that we should draw number of purchases x_{it} from $Poisson(\lambda)$, $\lambda \in (0, 1)$. Then, we set the prior distribution as $Gamma(2, 0.5)$, so that the expectation of λ generated from this distribution is 0.5.

7.3.1 *Fitness of Purchase Rates*

We use Mean Absolute Error (MAE) as the evaluation metric for the fitness of purchase rate. For each customer, we first compute the absolute difference between the estimated and actual number of purchase events for each time interval, and average

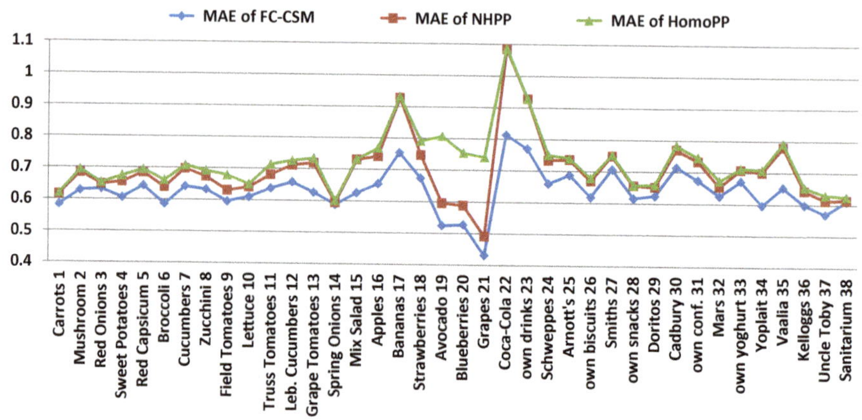

Fig. 7.3 MAE of purchase rates of 38 products for FC-CSM, NHPP and HomoPP

the value across T intervals. Then, the MAE for a product is the mean absolute difference obtained for $|A|$ customers, computed by:

$$\text{MAE} = \frac{\sum_{i=1}^{|A|}(\sum_{t=1}^{T} |\lambda_i(t) - x_{it}|)}{|A|T} \tag{7.20}$$

We compare the model with:

- Baseline HomoPP, which is a mixture of standard homogeneous Poisson processes;
- NHPP, which uses a mixture of Poisson processes with a polynomial and a periodic component in the intensity function, as described in Chap. 5.

Figure 7.3 shows that the MAE of our model is lower than the MAE of both HomoPP and NHPP for all 38 products. The detailed MAE values in this figure are provided in Table 7.3. The average MAE of our model across all products is 0.63, which is significantly lower than the average MAE of HomoPP, 0.73 ($p < 0.001$) and of NHPP, 0.70 ($p = 0.002$). On average, our model has decreased MAE of NHPP by 10%. We also notice that our model has achieved greater improvements for fruits (14.35%), soft drinks (21.72%) and yoghurt (13.93%), which jointly have an average improvement of 16.67%. For the other products, the average increase of fitness is 6.99%. The possible reason is that the changes of seasons have stronger impact on fruits, soft drinks and yoghurt than on the other products. Although NHPP can capture long-term behaviour patterns via the polynomial component, the dynamic partition of our model is more flexible and can generate a lower MAE for these products.

7.3.2 Impact of Promotions

We examine the impact of promotions via the correlation between average purchase rates and price. Negative correlation means that the purchase rate is higher when the price decreases, so smaller correlation values indicate a larger impact of promotions.

Table 7.3 Average MAE values across all customers for HomoPP, NHPP and FC-CSM; correlation values between the purchase rates computed by FC-CSM and price for all products

ID	Category	Product name	HomoPP	NHPP	FC-CSM	Correlation with price
1	Vegetables	Carrots	0.619	0.615	0.583	−0.422
2		Cup Mushrooms	0.694	0.684	0.628	0.086
3		Red Onions	0.653	0.648	0.633	0.028
4		Sweet Potatoes	0.674	0.655	0.606	−0.235
5		Red Capsicum	0.695	0.685	0.643	0.176
6		Broccoli	0.661	0.639	0.586	0.136
7		Cucumbers	0.709	0.698	0.642	−0.3
8		Zucchini	0.692	0.673	0.633	−0.036
9		Field Tomatoes	0.678	0.63	0.596	−0.503
10		Lettuce	0.651	0.641	0.61	−0.344
11		Truss Tomatoes	0.714	0.681	0.638	−0.544
12		Leb. Cucumbers	0.725	0.711	0.658	−0.323
13		Grape Tomatoes	0.734	0.719	0.626	−0.467
14		Spring Onions	0.600	0.595	0.589	−0.535
15		Mix Salad	0.732	0.731	0.626	−0.13
16	Fruits	Apples	0.768	0.742	0.655	−0.338
17		Bananas	0.932	0.928	0.754	0.13
18		Strawberries	0.792	0.747	0.674	−0.543
19		Avocado	0.807	0.596	0.525	−0.812
20		Blueberries	0.755	0.589	0.528	−0.817
21		Grapes	0.742	0.492	0.431	−0.832
22	Soft Drinks	Coca-Cola	1.088	1.085	0.813	−0.266
23		own drinks	0.929	0.928	0.77	−0.218
24		Schweppes	0.751	0.734	0.66	−0.45
25	Biscuits	Arnott's	0.741	0.737	0.688	0.093
26		own biscuits	0.679	0.669	0.619	−0.061
27	Snacks	Smiths	0.753	0.749	0.707	0.062
28		own snacks	0.658	0.656	0.616	−0.162
29		Doritos	0.659	0.652	0.624	−0.246
30	Confectionery	Cadbury	0.784	0.77	0.715	−0.025
31		own conf.	0.746	0.734	0.674	−0.099
32		Mars	0.673	0.654	0.625	−0.491
33	Yoghurt	own yoghurt	0.712	0.708	0.672	−0.031
34		Yoplait	0.708	0.7	0.597	0.009
35		Vaalia	0.792	0.78	0.653	0.007
36	Cereal	Kelloggs	0.654	0.643	0.599	−0.094
37		Uncle Toby	0.630	0.613	0.569	−0.078
38		Sanitarium	0.625	0.614	0.611	−0.232

The results are presented by the bar chart in Fig. 7.4. The correlations in this figure are also listed in the last column of Table 7.3. Among all 38 products, 20 correlation values are lower than −0.2, which means that the customers buying these products are more receptive to promotions.

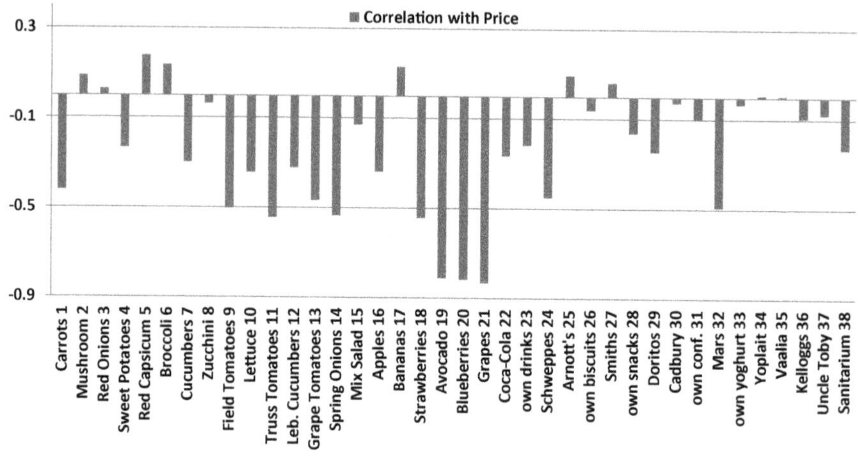

Fig. 7.4 Correlations between price and purchase rates of 38 products

At the category level, the average correlation of all fruits excluding bananas is strongly negative, −0.67. The vegetables also have low correlation values, −0.23 on average. The prices of fruits and vegetables generally have more significant changes than other products, due to the varying product availability. It makes sense that the price of these fresh products has significant impact on the purchase decisions. As for the other categories, the promotions of soft drinks and confectionery are generally more effective than of biscuits and snacks. Thus, the regular promotions with high discount rates of soft drinks may play an important role in attracting customers.

7.3.3 Trajectory of Purchase Behaviour Changes

Given the sequence of customer partitions, we first summarise the group and customer statistics, and then demonstrate the trajectories of customer partitions and compare individual purchase behaviour curves.

Statistics of Customer Groups

The total number of groups for different products over the observation period varies between 21 and 48. Figure 7.5 shows the histogram of total number of groups for each product. The x-axis is the bin of the histogram, where [0, 25] means the number of groups is less than or equal to 25. There are 13 out of 38 products having 26–30 groups during the one-year observation.

The average lifetime of a group is 2.15, which corresponds to 30.1 days, as the unit of time is 14-day. About 14% of the groups lasted for more than 8 weeks, which indicates that the partition changes at a moderate pace.

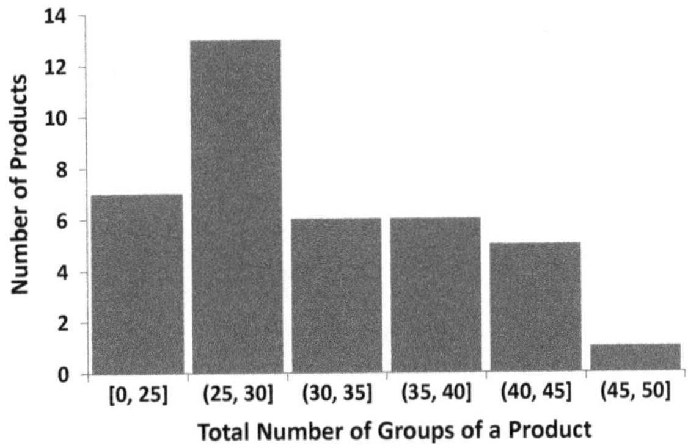

Fig. 7.5 Histogram of the total number of groups for each product

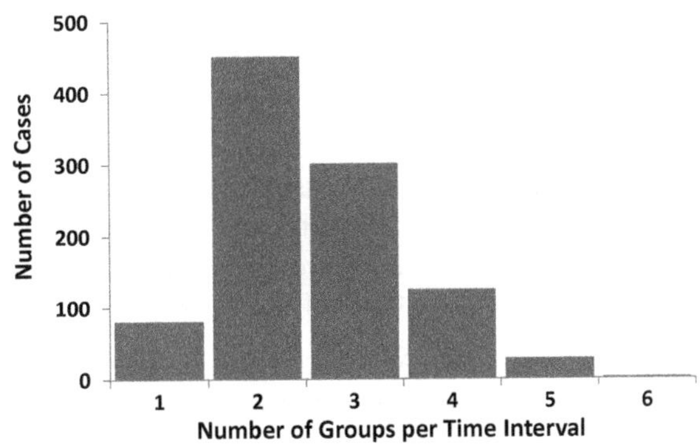

Fig. 7.6 Histogram of the number of groups per time interval (time unit: two weeks)

The number of groups at each time interval varies from 1 to 6, and the average value is 2.57. As for HomoPP and NHPP, their numbers of groups remain at 3 over time. The histogram of the number of groups per time interval of all products is shown in Fig. 7.6. It shows that 76.11% of time intervals have 2 or 3 groups.

We also illustrate the number of groups for two products (Vaalia yoghurt and avocado) and the average number of groups across all products over the observation period in Fig. 7.7. We notice that our model has improved the fitness of purchase rates of Vaalia and avocado by 19.33 and 13.49% when compared to NHPP based on Fig. 7.3. Given the improvements of fitness and the varying numbers of groups for these two products (as shown in Fig. 7.7), it clearly demonstrates the advantage of using varying number of groups in FC-CSM rather than fixing the number of groups. In terms of the average number of groups across all products, we find that the value is slightly higher in mid-year (around May and June) than the other periods of time, which may be related with the start of the health program.

Trajectory of Customer Partitions

We analyse the trajectory of customer partitions for each product and illustrate the duration, size and purchase rate of all customer groups using bubble plots shown in Fig. 7.8. Each bubble represents a customer group with the group ID in it, and its

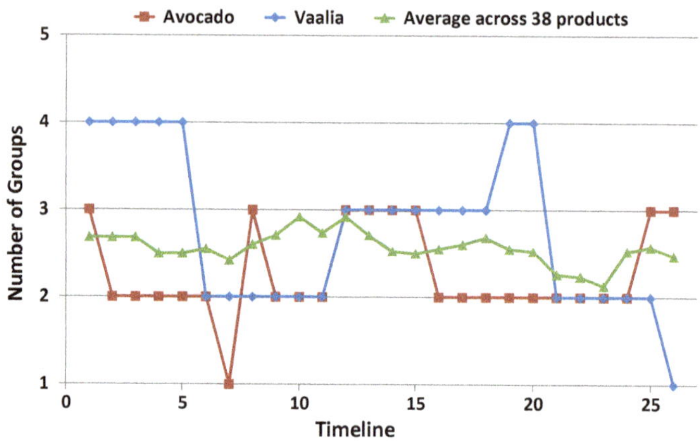

Fig. 7.7 Number of groups for Vaalia yoghurt, avocado and the average of all products over the one-year observation (time unit: two weeks)

size shows the proportion of customers in that group. Since the customer groups at a certain time interval are exclusive and non-overlapping, each customer is in one and only one of the group at any time. We select four representative plots as examples, corresponding to three types of behaviour patterns:

(1) stable behaviour (Arnott's biscuits, top);
(2) behaviour strongly affected by seasonal availability (grapes, second);
(3) varying behaviour, with customers dynamically shifting across groups (avocado, third; Vaalia yoghurt, bottom).

The top plot for Arnott's biscuits shows that there are two different values of λ. About 25% of customers had λ above 1.5, while the other 75% of customers had λ below 0.7. We notice that there were not many shifts between the two levels, which means that the purchase behaviour was relatively stable for this product.

The second plot for grapes shows the impact of the seasonal product availability on purchase behaviour. There is a major group steadily including about 90% of the customers. Overall, the λ curve of this group decreased significantly at $t = 10$ (highlighted by the dashed line) and formed a U shape, which is the result of the high price and low supply in mid-year, when it is winter in Australia. Our model has also identified minor groups of customers, such as Group 2. In contrast, mixture models would have included these customers in the major group. The minor groups identified by our model can be used by stakeholders to discover customers who differ from the majority but are worth noting.

The third plot for avocado illustrates different responses to promotions from different customers. There is only one major group at the first 7 time intervals. The λ values increased from $t = 8$ (highlighted by the first dashed line) for all customers and the major group split into two streams at that time, due to different responses to promotions. About 30% of the customers were more receptive to promotions and had λ values between 1.5 and 2.2. The remaining customers were less receptive to

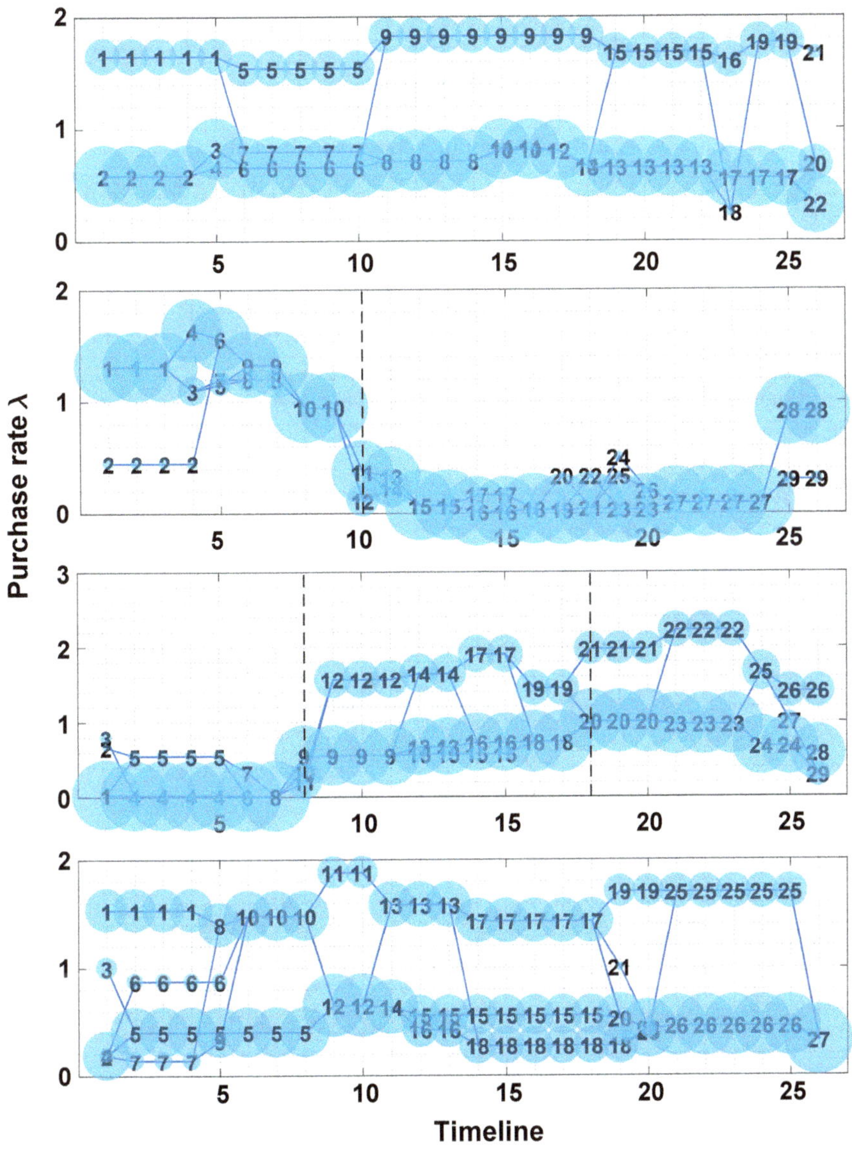

Fig. 7.8 Trajectory of customer partitions based on the number of purchases (time unit: two weeks). The subplots are for Arnott's biscuits (top), grapes (second), avocado (third) and Vaalia yoghurt (bottom). The dashed lines highlight the time intervals discussed in the text

promotions and had λ values between 0.5 and 1. Some customers shifted from one level to the other. For example, at $t = 18$ (highlighted by the second dashed line), most of the customers in Group 19 increased their purchase rates and formed Group 21, while a small number of customers decreased λ and joined Group 20.

The last plot for Vaalia yoghurt demonstrates varying behaviour and customers shifting across groups. The number of groups of Vaalia over time has been presented in Fig. 7.7. We notice that this product has more customer groups at some time intervals than the previous three products. For instance, there are 4 customer groups from $t = 1$ to $t = 5$, which includes minor groups such as Group 6. The customers in Group 6 were from Group 2, and they increased their purchase rates and joined Group 10. Then, the customers in Group 10 split into two streams at $t = 9$, with higher and lower purchase rates. Our model can support stakeholders to identify customers in interesting groups for further analysis, such as Group 6 and Group 11 who increased their purchase rates more than the others.

Trajectory of Individual Purchase Rate

For an individual customer, the purchase rate curve $\lambda_i(t)$ can track how the customer changes behaviour and shifts across groups, which supports the fine-grained analysis at individual level.

We present three sets of customers with contrasting behaviour in Fig. 7.9 (the group index has been omitted for clarity). When the lines with different colours join

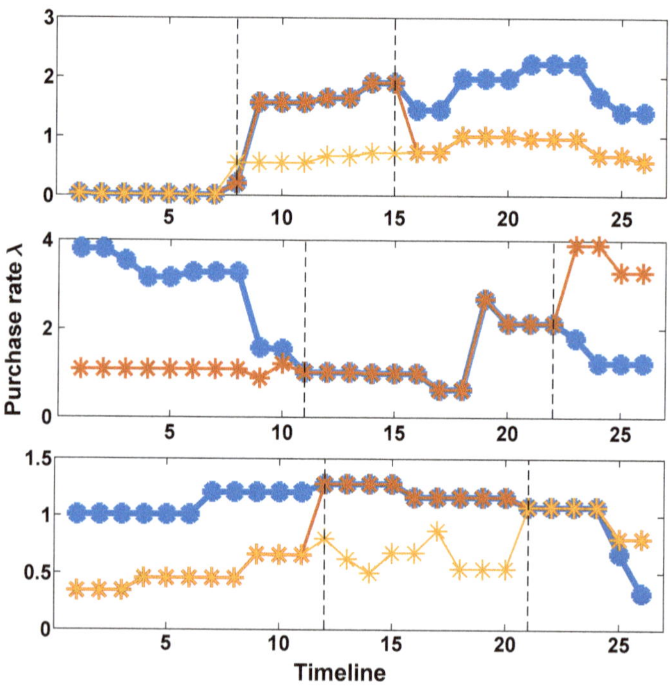

Fig. 7.9 Individual purchase rate curves of different customers for avocado (top), Coca-Cola (middle) and broccoli (bottom) (time unit: two weeks). Each plot demonstrates contrasting behaviour of multiple customers for one product, and each line corresponds to one customer. The dashed lines highlight the time intervals discussed in the text

together, it means that the corresponding customers are in the same group during that period of time. The dashed lines highlight the time intervals with significant behaviour changes.

The top plot includes three customers buying avocado. The turning points of their purchase curves were at $t = 8$ and $t = 15$ (highlighted by the dashed lines), when the price changed substantially. For example, at $t = 15$, customer 2 (red) decreased λ and joined customer 3 (yellow), whereas customer 1 (blue) maintained her high purchase rate until the end.

The middle plot displays two customers buying Coca-Cola who were in the same group for 22 weeks. However, customer 1 (blue) used to have significantly larger λ than customer 2 (red) before $t = 11$. They split again after $t = 22$, when customer 1 decreased λ gradually, whereas customer 2 increased λ abruptly. It shows that customers with similar total number of purchases can have contrasting behaviour changes.

The bottom plot contains three customers buying broccoli. Customer 2 (red) and customer 3 (yellow) were in the same group for 17 time intervals, which were before $t = 12$ and after $t = 20$. We find customer 2 increased her purchase rate and joined customer 1 (blue) at $t = 12$, earlier than customer 3, who joined the other two customers at $t = 21$. All of them remained in the same group for 8 weeks, until customer 1 decreased purchase rate before the end of the year.

The trajectory of individual purchase rate facilitates the comparison of customers, so that we can identify their similarities and differences. Moreover, the bubble plots in Fig. 7.8 actually combine the individual purchase rate curves of all customers, where the solid lines connecting bubbles mean that at least one customer takes that path.

7.4 Summary

In this chapter, we propose a Bayesian nonparametric customer segmentation model FC-CSM, which can track the evolution of customer groups using a fragmentation-coagulation process. The FC-CSM can split groups when customer behaviour diverges and merge multiple groups when their behaviour becomes similar. The model outputs the sequence of customer partitions over time, which can capture the dynamics of customer behaviour and address the group-level behaviour shifts. Moreover, the FC-CSM is also flexible and can be applied to various types of behaviour data.

Our case study uses the supermarket transaction records of 38 products. We analyse the evolution of customer groups and how it is affected by product promotions and seasonal changes. The case study shows that FC-CSM is more accurate than a mixture of Poisson processes on estimating purchase events. We find that customers were more receptive to promotions of fresh fruits, soft drinks and confectionery. The sequence of customer partitions shows that for products with strong impact of promotions, the customer groups tend to split during promotions and merge when the price stabilises. In addition to group trajectory, FC-CSM also shows the trajectories of individual customers, which track how they shift across groups and how the customers with similar number of purchases may have contrasting behaviour patterns.

Our model FC-CSM offers a strong and sensitive tool for analysing customer purchase behaviour and tracking the effects of promotions and seasonal campaigns. For example, our model can be used to find customers who were not receptive to previous strategies, so that the stakeholders can adjust their strategies to attract them. We believe that the information provided by FC-CSM can be used by stakeholders in order to understand customer behaviour changes, identify customer groups, and also optimise the timing and focus of their promotion campaigns.

Comparing FC-CSM in this chapter and the mixture of NHPP model from Chap. 5, we can see that they have different strengths. Although FC-CSM is more flexible and accurate than NHPP when tracking the "group-level behaviour shifts", NHPP is particularly useful when the users are interested in detecting and differentiating long-term and short-term behaviour patterns of different customer groups.

References

1. Bucklin RE, Gupta S, Siddarth S (1998) Determining segmentation in sales response across consumer purchase behaviors,. J Mark Res, 189–197
2. Bertoin J (2006) Random fragmentation and coagulation processes. vol 102. Cambridge University Press
3. Luo L, Li B, Koprinska I, Berkovsky S, Chen F (2016) Discovering temporal purchase patterns with different responses to promotions. In: Proceedings of the 25th ACM international conference on information and knowledge management. ACM, pp 2197–2202
4. Luo L, Li B, Berkovsky S, Koprinska I, Chen F (2017) Online engagement for a healthier you: a case study of web-based supermarket health program. In: Proceedings of the 26th international conference on world wide web companion. International world wide web conferences steering committee, pp 1053–1061
5. Iwata T, Shah A, Ghahramani Z (2013) Discovering latent influence in online social activities via shared cascade Poisson processes. In: Proceedings of the 19th ACM SIGKDD international conference on knowledge discovery and data mining. ACM, pp 266–274
6. Blei DM, Lafferty JD (2006) Dynamic topic models. In: Proceedings of the 23rd international conference on machine learning. ACM, pp 113–120
7. Lan D, Buntine W, Jin H, Chen C (2012) Sequential latent Dirichlet allocation. Knowl Inf Syst 31(3):475–503
8. Spiliopoulou M, Ntoutsi I, Theodoridis Y, Schult R (2006). MONIC: modeling and monitoring cluster transitions. In: Proceedings of the 12th ACM SIGKDD international conference on knowledge discovery and data mining, New York, NY, USA. ACM, pp 706–711
9. Xing EP, Sohn K-A (2007) Hidden Markov Dirichlet process: modeling genetic inference in open ancestral space. Bayesian Anal 2(3):501–527
10. Jasra A, Holmes CC, Stephens DA (2005) Markov chain Monte Carlo methods and the label switching problem in Bayesian mixture modeling. Statist Sci, 50–67
11. Eguiluz VM, Zimmermann MG (2000) Transmission of information and herd behavior: an application to financial markets. Phys Rev Lett 85(26):5659–5662
12. Tadić B, Rodgers GJ (2010) Modelling conflicts with cluster dynamics in networks. Phys A: Stat Mech Appl 389(23): 5495–5502
13. Teh YW (2011) Charles Blundell, and Lloyd Elliott, "Modelling genetic variations using fragmentation-coagulation processes. Adv Neural Inf Process Syst, pp. 819–827

14. Elliott L, Teh YW (2012) Scalable imputation of genetic data with a discrete fragmentation-coagulation process. Adv Neural Inf Process Syst, 2852–2860
15. Pitman J (2002) Combinatorial stochastic processes. Technical Report. no 621
16. Heinrich G (2008) Parameter estimation for text analysis. Technical Report
17. Bernardo JM, Smith AFM (1994) Bayesian theory. Wiley
18. Sheldon M (1996) Ross, stochastic processes, vol 2. Wiley, New York
19. Frühwirth-Schnatter S (1994) Data augmentation and dynamic linear models. J Time Ser Anal 15(2):183–202

Chapter 8
Conclusions and Future Work

8.1 Conclusions

In this thesis, we have proposed temporal modelling techniques for customer behaviour analysis, which are based on temporal collaborative filtering and stochastic processes. We have conducted four case studies on a large-scale real-world supermarket health program. The datasets used in our case studies include supermarket transaction records, surveys of customers' demographic and health information, and activity logs of the supporting website. Our case studies show that the proposed temporal modelling techniques can accurately track dynamic customer behaviour, detect purchase behaviour patterns of different customers and identify important factors that impact customer behaviour. We have provided answers to the following research questions:

Q1: When do customers buy the products?
Q2: How do customers change their purchase behaviour over time?
Q3: What are the temporal patterns of customer purchase behaviour?
Q4: How would different types of customers respond to marketing strategies?

Our main contributions can be summarised as follows:

(1) We propose innovative temporal modelling techniques based on collaborative filtering model RMGM-OT and Poisson processes to track the customer behaviour changes.
(2) We discover long-term and short-term purchase behaviour patterns due to various factors, such as seasonal effects and drifting customer preferences over time using a mixture of Non-Homogeneous Poisson Processes (NHPP).
(3) We segment customers into groups based on demographic and behavioural attributes, which can help people to identify and compare different types of customers.
(4) We propose a dynamic customer segmentation model FC-CSM, based on the Bayesian nonparametric fragmentation-coagulation process to track the evolu-

L. Luo, *Temporal Modelling of Customer Behaviour*,
Springer Theses, https://doi.org/10.1007/978-3-030-18289-2_8

tion of customer groups, including splitting and merging of groups, and monitor the dynamic customer behaviour at group and individual levels.
(5) We evaluate the impact of: (i) various promotions, (ii) a health program, and (iii) the supporting website of the health program, on different types of customers.

In more detail, we proposed a temporal model to explore customer preferences for various food categories and track customer preferences drifting over time. We evaluated the impact of our supermarket health program on different types of customers. To identify different types of customers, we segmented customers based on demographic and health attributes such as age, gender, living arrangements and BMI. Our case study showed that the health program successfully motivated customers to purchase healthier food. Specifically, the segment-wise impact scores showed that the health program affected different types of customers to different degrees. We found that female customers, younger customers, customers who live with their family and obese customers were more receptive to the health program than their counterparts.

After that, we proposed another purchase behaviour model based on a mixture of Non-Homogeneous Poisson Processes (NHPP). In our NHPP model, we used an intensity function with a polynomial and a periodic component to capture the latent long-term and short-term purchase behaviour patterns shared by customers. Our mixture model can segment customers based on their purchase behaviour patterns. We conducted a case study on the purchase records of food categories. The results showed that our NHPP model outperformed homogeneous Poisson processes on estimating the occurrence of purchase events. We found that the distribution of long-term patterns for a product is often determined by whether the demand of this product is sensitive to seasonal effects or personal preference changes. As for short-term periodic patterns, the main factors are regular promotions and whether the product is easy to stockpile at home. We also compared customer responses to promotions using the correlation between purchase rates and price, and identified customers who are receptive to price reductions.

Based on the NHPP model, we evaluated the impact of our supermarket health program and its supporting website, using transaction records and activity logs of the supporting website. We tracked individual purchase behaviour changes and conducted a fine-grained analysis across groups of customers and products. We designed four criteria for partitioning the customers by their types and levels of interactions with the program website. We found that the active web users, active diary users, customers with 2 valid weigh-ins and customers with $\geq$3kg weight loss have been motivated to a greater extent to buy healthier food. Based on the impact scores, these customers had larger increases in their purchase rates for some fresh products, especially fruits. Among the four partition methods, we found the program impact was more pronounced on the participants interacting with the program website than the rest, especially when the criterion considers higher levels of interactions with the program website, such as using the weight tracker. Our model allows uncovering the hidden dependencies and impact of the health program, which may be overlooked by the traditional overall analysis methods without behavioural tracking.

To track the evolution of customer groups, we proposed a Bayesian customer segmentation model via the fragmentation-coagulation process, named as FC-CSM. FC-CSM can address the group-level behaviour shifts by splitting and merging groups. More specifically, the model splits groups when customer behaviour diverges and merges groups when customer behaviour becomes similar. Our model is flexible to use, which can automatically adjust the number of groups based on the dataset. Our case study on the supermarket transaction records showed that FC-CSM was more accurate than the mixture of NHPP on estimating the purchase events. Given the customer partition sequence, we analysed the evolution of customer groups of popular products, and how customer groups were affected by product promotions and seasonal changes. We found that for products that are highly impacted by promotions, the customer groups tended to split during promotions and merged when the price stabilised. In addition to the analysis at group level, we also tracked the customer behaviour at individual level via FC-CSM, and showed that the customers with similar number of purchases may have contrasting behaviour patterns.

8.2 Future Work

Our contributions represent a promising step towards a better understanding of customer behaviour via accurate and flexible temporal models. There are various ways to extend our work, both the theoretical and applied aspects of it.

From a *theoretical* perspective, a potential research direction is to investigate how a promoted product influences the purchase behaviour of a similar product. We could consider multiple stochastic processes and their interactions simultaneously.

Furthermore, we could model the customer purchase behaviour at the basket level. This allows comparing proportions of different products in one shopping trip, instead of examining the fluctuations of the purchase rates of different products separately. The basket-level modelling can provide more insights to evaluate the purchase behaviour changes and the impact of health programs.

We could also strengthen our temporal modelling framework of customer behaviour by extending it from tracking behaviour changes and measuring program impact to predicting future customer purchase preferences.

Another research direction for future work is the causality between purchase behaviour and different factors. In this thesis, we evaluated the customer's receptiveness to the health program by comparing their purchase behaviour changes. To identify and understand the causal relationship, it requires the transaction records of customers who had similar pre-intervention purchase behaviour but did not attend the health program.

One of the limitations of our case studies is that we do not have ground truth or domain experts to verify the customer segmentation and behaviour patterns obtained from the proposed models. We will consider to validate and reinforce our results by qualitative studies from other disciplines such as social science and marketing.

From an *applied* perspective, the future work can explore how to process the model outputs with a systematic and comprehensive procedure, so that the models are capable of providing concrete guidelines to optimise marketing strategies. For example, the stakeholders should be provided with specific analytical results based on their needs, such as identifying interesting groups based on customised criteria, and generating a list of target customers for marketing campaigns. Moreover, our models can be easily applied to different programs with social values to benefit the wider society.

Considering the advancement of data mining and user behaviour analytics, we can adapt our temporal modelling techniques of supermarket customer behaviour to other areas, such as online shopping, energy usage, activities on social networks and educational systems. For example, we can extend FC-CSM to capture the evolution of the user groups on social networks. Another idea is to use the NHPP model to capture the student behaviour patterns in online learning platforms and detect the critical factors affecting the learning performance.

Author Biography

Dr Ling Luo is currently an Associate Lecturer at the School of Computer Science, University of Technology Sydney, Australia. Before this, Ling was a Postdoctoral Research Fellow in Data Analytics at Data61 (formerly NICTA), CSIRO, Sydney.

Ling received Ph.D. from the University of Sydney in 2017. Ling's supervisors were Dr Irena Koprinska from the School of Computer Science (formerly School of Information Technologies), the University of Sydney, and Dr Bin Li from the School of Computer Science, Fudan University, CSIRO. Prior to her Ph.D., Ling studied at the University of Sydney, receiving a Bachelor of Engineering degree in Software Engineering, with Honours Class I, in 2012. Ling has been awarded more than 10 academic awards and prizes, including the Google Ph.D. Fellowship in Machine Learning in 2017, which is awarded to the "most promising young academics in the world" doing "exceptional research in computer science".

This book is the result of Ling's Ph.D. thesis in computer science, which explores modelling dynamic customer behaviour with advanced machine learning techniques. Ling's research interests include temporal modelling, user behaviour analysis, stochastic processes and Bayesian nonparametrics. Ling has published at prestigious conferences and journals in data mining, machine learning and artificial intelligence. Among Ling's papers, the work published at the International Joint Conference on Artificial Intelligence (IJCAI) in 2017 was selected for the press release opening of the conference "highlighting new and exciting applications of AI research".

The full lists of Ling's publications and achievements are available at https://sites.google.com/site/lingluoau/home.

L. Luo, *Temporal Modelling of Customer Behaviour*,
Springer Theses, https://doi.org/10.1007/978-3-030-18289-2

CPSIA information can be obtained
at www.ICGtesting.com
Printed in the USA
LVHW051937160519
618116LV00002B/6/P